Strategic Research on Construction and Promotion of China's Intelligent Cities

Editor-in-Chief

Yunhe Pan, Chinese Academy of Engineering, Beijing, China

This book series is the first in China on "Intelligent City" research, with systematic and thorough contributions from more than 200 Chinese experts including 47 academicians of the Chinese Academy of Engineering (CAE) in related fields. The book series is co-published with Zhejiang University Press, Hangzhou, China and consists of 13 volumes as planned, including one general report and 12 sector reports. In 2010, CAE conducted a research on the development of "smart cities" and concluded that urban development in China has reached a crucial turning point. Therefore, CAE kicked off the key consultancy research project on "Strategic Research on Construction and Promotion of China's Intelligent Cities", on which this book series is based. Firsthand and research results, surveys and analysis are provided on almost every aspect of urban development and smart cities in this series. Representing the highest level of research in this field in China, the book series will offer an authoritative reference resource for international readers, helping them to understand intelligent city construction in China, a movement expected to be highly influential around the globe.

More information about this series at http://www.springer.com/series/15953

Zhiqiang Wu

Intelligent City Evaluation System

ZHEJIANG UNIVERSITY PRESS
浙江大学出版社

Springer

Zhiqiang Wu
Tongji University
Shanghai
China

ISSN 2522-8943 ISSN 2522-8951 (electronic)
Strategic Research on Construction and Promotion of China's Intelligent Cities
ISBN 978-981-13-3865-6 ISBN 978-981-10-5939-1 (eBook)
https://doi.org/10.1007/978-981-10-5939-1

Jointly published with the Zhejiang University Press, Hangzhou, China

The print edition is not for sale in China Mainland. Customers from China Mainland please order the print book from: Zhejiang University Press

Printed on acid-free paper

This Springer imprint is published by the registered company Springer Nature Singapore Pte Ltd.
part of Springer Nature
The registered company address is: 152 Beach Road, #21-01/04 Gateway East, Singapore 189721, Singapore

Foreword

In 2008, IBM proposed the concept of "Smarter Planet", in which "Smart City" was one of its components, mainly referred to 3I, namely, instrumented, interconnected, and intelligent, and the goal was to implement the company's "solutions", such as smart transportation, medical, government services, monitoring, grid, water, and other items.

In early 2009, US President Barack Obama publicly acknowledged IBM's "Smarter Planet" concept. In December 2012, the Global Trends 2030, published by the National Intelligence Council, noted that the four most influential technologies for global economic development were information technology, automation & manufacturing technology, resource technology, and health technology, in which "smart city" was one of the information technology contents. Envisioning 2030: US Strategy for the Coming Technology Revolution report pointed out that the world was on the cusp of the next major technological change, the "third industrial revolution" represented by manufacturing technology, new energy, and smart city would have an important influence on shaping future political, economic, and social development trends.

In May 2011, after the implementation of the "i2010" Strategy, the EU Net! Works forum introduced Smart Cities Applications and Requirements white paper, emphasizing low carbon, environmental protection, and green development. After that, the EU said it would set "Smart City" as the key development content of the Eighth Framework Programme (FP8).

In August 2009, in the Smarter Planet to Win in China plan, IBM tailored for the Chinese six intelligent solutions: "Intelligent Power", "Intelligent Medical", "Intelligent City", "Intelligent Traffic", "Intelligent Supply Chain", and "Intelligent Banking". In 2009, "Smart City" spread in succession at all levels in China, as till September 2013, there were a total of 311 cities under construction or planning to build smart city in China.

In 2010, Chinese Academy of Engineering carried out research on the "smart city" construction. It considered that the current urban development in China had reached a key transition period, but since national conditions were different, there were still some problems in "Smart City" construction.

To this end, in February 2012, the Chinese Academy of Engineering launched a major consulting research project "China Intelligent City Construction and Promotion Strategy Research". Since the project started, many city leaders and scholars have shown a keen interest and expected to devote to the research and practice of intelligent city construction. With the strong support of people from all walks of life and the efforts of academicians and experts of the "China Intelligent City Construction and Promotion Strategy Research" project group of the Chinese Academy of Engineering, we have combined three research efforts: experts of the relevant national ministries (such as the National Development and Reform Commission, the Ministry of Industry and Information Technology of the People's Republic of China, the Ministry of Housing and Urban-Rural Development of the People's Republic of China), experts of typical cities (such as Beijing, Wuhan, Xi'an, Shanghai, Ningbo, etc.), 47 academicians and more than 180 experts of the Department of Information and Electronic Engineering of the Chinese Academy of Engineering, the Department of Energy and Mining Engineering, the Department of Environment and Textile Engineering, the Department of Engineering Management and the Department of Civil Engineering, Water Conservancy and Construction Engineering, and other departments. The research project was divided into 13 task groups, involving urban infrastructure, information, industry, management, and so on. In addition, the project also has a comprehensive group, whose main task is to comprehensively formulate the integrated volume of the book series *Strategic Research on Construction and Promotion of China's Intelligent Cities* on the basis of the results of 13 task groups.

After more than 2 years, the research team has formed a number of research results and research comprehensive report by carrying out inspection and investigation in the site, carrying out the forums and exchanges with experts and scholars at home and abroad, having informal discussion with the national authorities and local authorities responsible comrades, and team research and analysis itself, and so on. In the study, we put forward that it will be more suitable for China's national conditions to carry out intelligent city (iCity) construction and promotion in China. The intelligent city construction will become an accelerant of deepening the system reform and development, and become a strong starting point for the economic and social development and the realization of "China Dream" in China.

Beijing, China   Yunhe Pan
                                                                              Chinese Academy of Engineering

Preface

The book series *Strategic Research on Construction and Promotion of China's Intelligent Cities* is compiled and published with the works of 47 academicians and over 180 experts on the basis of the research achievements obtained after over 2 years of in-depth investigation, research, and analysis and the study on China's Intelligent City Construction and Promotion Strategy, which is one of the major consulting and research projects conducted by the Chinese Academy of Engineering after revision in accordance with publishing requirements. The Book Series, consisting of one comprehensive volume and 13 sub-volumes, have been published in succession by Zhejiang University Press. The comprehensive volume mainly discusses how to conduct the intelligent city construction and promotion with Chinese characteristics in the intelligent urbanization development of our future cities, and the sub-volumes focus on the construction and promotion of intelligent city in terms of urban economy, science, culture, management & education, spatial organization pattern of cities, intelligent transportation & logistics, intelligent grid & energy network, intelligent manufacturing & design, knowledge center & information processing, intelligent information network, intelligent building & smart home, intelligent medical & health care, urban security, urban environment, intelligent business & finance, intelligent city's time & space information infrastructure, intelligent city's evaluation indicator system, etc.

As a consultant of "Strategic Research on Construction and Promotion of China's Intelligent Cities" project group, I have participated in several research meetings of the project group and put forward some "humble opinions". Overall, I think, under the leadership of the project group leader—academician Pan Yunhe, "Strategic Research on Construction and Promotion of China's Intelligent Cities" has made significant progress, which is mainly shown in the following aspects.

Since 1990s, the world entered the era of information technology, and the city has gradually developed from the traditional binary space to the ternary space. The first metaspace mentioned refers to a physical space (P), which consists of physical environment of the city and the urban physical; the second metaspace refers to a human social space (H), that is, the human decision-making and social interaction space; the third metaspace refers to a cyberspace (C), that is, the "network information" space composed by computer and Internet. The city intelligence is the development trend of cities throughout the world, even though the development stages of cities in each country are different, and the contents are different. At present, the "smart city" construction proposed at domestic and abroad is mainly focused on the building of the third metaspace, and the city intelligence of our country should be "ternary space" to coordinate with each other so that planning and industry, life, and social and social public service could be mutual cooperation and promotion, and it should be beyond the existing e-government, digital city, network city, and smart city construction concept.

The new technological revolution will promote the arrival of urban intelligence era. Nowadays, about the new technological revolution, authors hold varying viewpoints: there are "the second economy", "the third industrial revolution", "industry 4.0" "the fifth industrial revolution", and other concepts. And when it comes to the city, the new technological revolution is characterized by integrating a new generation of sensor technology, Internet technology, big data technology, and engineering technical knowledge into the city's systems to upgrade the quality of urban construction, urban economy, urban management, and public service, so as to embrace a new era of urban intelligence development. If China's urbanization and the new technological revolution are organically linked together, it can not only promote the benign and healthy development of China's urban intelligence process but also promote the birth of more new technologies. China shall undoubtedly actively participate in this process and make a greater contribution to the development of the world's economy, science, and technology.

It has been repeatedly deliberated by the project group to use "Intelligent City" (iCity) to replace "Smart City". The reasons are as follows: first of all, the western developed countries have completed the urbanization, industrialization, and agricultural modernization, and the main tasks of the smart city they refer to are limited to the intelligence of government management and service, and the administrative functions of their city managers are much narrower than those of city mayors in China; second, currently, China is in the simultaneous development stage of industrialization, informatization, urbanization, and agricultural modernization; the confusions and problems it encounters have uniqueness in quality and quantity, so China's urban intelligence development path must be different from the Europe and the United States, and it will be difficult to solve many development problems which Chinese cities confront by only interpreting the smart city from the perspective of developed countries and moving this concept to China. Thus, the project group decided to use the term "Intelligent City" (iCity) and expected this term would be more in line with China's national conditions.

The construction and promotion of intelligent cities have far-reaching significance to China's current economic and social development. The construction and promotion of intelligent cities are just located in the cross point of "Four Modernizations", and its significance is mainly embodied in the following aspects. First, it can be used as the basic platform for the simultaneous development of "Four Modernizations" and become a focal point of China's economic and social development to avoid the "middle-income trap", so as to create a new urbanization development path with Chinese characteristics. Second, by putting the intelligent city as an important basis (point), it can promote the development of "One Belt, One Road" (line) and new area (plane) and constitute a reasonable development layout of "point, line and plane". Third, it is conducive to promote the structure upgrading and transformation of manufacturing and its service industry, to achieve the transformation from the urban industry to the intensive pattern so as to slow down the material growth, accelerate the value growth, and improve the added value, and it is conducive to the usage and integration of a variety of e-commerce, big data, cloud computing, and Internet of Things technologies to achieve the "broadband, pan, mobile, integration, security, green" development of information and network technology, promote the improvement of urban industry efficiency, form the new factors of production and new formats, and create new conditions for entrepreneurship and employment. Fourth, it developed from the simple and linear decision-making based on limited information to the networked and optimized decision-making based on urban comprehensive system information, so as to help the government improve the level of urban management services and promote the in-depth urban administrative system reform and development. Fifth, it can use new technologies to optimize and improve the planning of urban construction, roads, transportation, energy, resources, environment, etc., to increase the utilization efficiency of elements, and further protect, inherit, develop, and sublimate the urban history, landforms, local culture, etc., and achieve the change of the public health management from the concept to reality, and so on. Sixth, it can find and cultivate a number of urban planners, management experts, high-level scientists, data science and security experts, engineering and technical experts, etc. to adapt to the new technological revolution trends, learn from past experience and lessons, and pay attention to the renovation during intelligent city operation and maintenance, and it can focus on cultivating a large base number of operation and maintenance engineers and technical staff of urban functions to both understand the theory and practice, achieving the gradual transformation from relying on the demographic dividend to relying on knowledge and talent dividends so as to support health, sustainable development of China's urban intelligence.

To sum up, the book series *Strategic Research on Construction and Promotion of China's Intelligent Cities* has rich content and clear views, and the proposed development goals, ways, strategies, and recommendations are reasonable and operational. I think this series of books is the literature of urban management innovation and development research with high reference value, and they have important theoretical significance and practical value to the development of new urbanization in China. I believe that readers of all sectors will get a lot of new inspiration and gain after reading them, and the book will inspire the enthusiasm of everyone to participate in the construction of intelligent city, so as to put forward more thinking and unique insights.

China is a developing country with long history and big agricultural population, and is committed to the economic society to be developed in good and fast and commercial manner and the construction of new urbanization. I am convinced that the publication of the book series *Strategic Research on Construction and Promotion of China's Intelligent Cities* will play an active and positive role in promoting this. Let us strive and work together for the realization of the great "Chinese dream"!

Shanghai, China   Zhiqiang Wu
January 2015

Acknowledgements

Xu Xingjing	Shanghai Tongji Urban Planning & Design Institute	Assistant Researcher
Lv Hui	Technische Universität Wien	Assistant Researcher
Kong Lingyu	College of Architecture and Urban Planning, Tongji University	Master
Ye Qiming	College of Architecture and Urban Planning, Tongji University	Master
Teng Yuwei	College of Architecture and Urban Planning, Tongji University	Master
Jing Han	Shanghai Tongji Urban Planning & Design Institute	Engineer
Lu Rongli	China Academy of Urban Planning & Design	Engineer
Sheng Xuefeng	Shanghai Pudong Smart City Research Institute	Executive Dean
Bo Yang	Shanghai Tongji Urban Planning & Design Institute	Engineer
Yu Jing	College of Architecture and Urban Planning, Tongji University	Master
Liu Chaohui	Tongji University Intelligent Urbanization Co-creation Center	Doctor
Cui Binghong	Tongji University Strategy Development Institute	Researcher
Yang Xiu	College of Architecture and Urban Planning, Tongji University	Doctoral Student

Proofreaders: Wang Zhihan, Gu Ruoyao

Contents

Chapter 1
Global Plan of Intelligent City

Abstract In the past more than a century, the world has experienced a rapid urbanization process. In 1900, only 13% of the world's population lived in the city, and by 2008 the proportion reached 50% for the first time. By the end of 2011, the urban population in China reached 691 million, accounting for 51.27% of the total population, and the urbanization rate of China exceeded 50% historically. This indicates that China and the world have officially entered the urban society. Massive urbanization creates opportunities on one hand and poses challenges on the other hand. The extensive development mode coupled with the uneven distribution of resources threatens urban ecosystem and makes the cities unsustainable. Meanwhile, a new revolution of new information communication technologies represented by Internet of Things, cloud computing, big data and next generation mobile communication technology has brought new perspectives and technical bases for the solution of traditional urban problems. In November 2008, IBM released the theme report of Smarter Planet: the Next Generation of Leaders Agenda in New York. Since then, the "smart city" concept was accepted gradually. The development of intelligent city was considered to help promoting the sustainable development of economy, society and environment, resource coordination in the city, alleviating the "big city disease", improving the quality of urbanization, so it quickly became an important fulcrum to meet the challenges of the city in the world.

Keywords Urban society · ICT · Smart city

1 A Global Review on Intelligent City Development

In the past more than a century, the world has experienced a rapid urbanization process.

In 1900, only 13% of the world's population lived in the city, and by 2008 the proportion reached 50% for the first time. By the end of 2011, the urban population in China reached 691 million, accounting for 51.27% of the total population, and the urbanization rate of China exceeded 50% historically. This indicates that China

© Springer Nature Singapore Pte Ltd. and Zhejiang University Press 2018
Z. Wu, *Intelligent City Evaluation System*, Strategic Research
on Construction and Promotion of China's Intelligent Cities,
https://doi.org/10.1007/978-981-10-5939-1_1

and the world have officially entered the urban society, the major changes of social demographic reflected by it will have a profound impact on the development of China and the world.

As a regional or national center area, the city highly gathers a lot of money, technology, talent, information and other material and social resources, and can greatly improve the efficiency of economic activities, form a good social environment and a variety of material and cultural life, and have an outstanding contribution on improving the social economic development level and the civilization level of a country or region. However, a large amount of rapid expansions of the city also have brought a lot of ordeal to the human society. The extensive development mode is becoming increasingly unsustainable, the environmental ecological binding force is increasing day by day, and the urban development is under increasing external pressure; the uneven distribution of resources, the unreasonable urban expansion, and the lag of public service level make the urban development face various internal pressures. The new population social structure presents new demands and challenges to traditional urban development and governance models.

Meanwhile, a new revolution of new information communication technologies represented by Internet of Things, cloud computing, big data and next generation mobile communication technology is becoming the focus of global social and economic development in the post-financial era.

The Internet of Things is another wave of information industry after the computer, the Internet and mobile communication network. Through interconnecting all the items with the network, the Internet of Things achieves intelligent identification and management, improving the "perception" ability of the city. The cloud computing technology has powerful data processing capabilities, and improves the intelligent processing capacity of the information. With the arrival of the cloud era, the big data is also attracting more and more attention. Reasonable and effective mining and processing big data will provide a solid and reliable basis for urban development.

The constant emergence and maturity of new technologies have brought new perspectives and technical bases for the solution of traditional urban problems. On the one hand, the industry development led by information communication technology can form a new round of urban development momentum to promote urban economic development; on the other hand, the technology itself can also provide a better technology for the diagnosis and analysis of urban problems. Through mining, processing and analyzing massive city information, it is conducive to find the optimal development path for each city to form the best input-output ratio.

In November 2008, under the background of global financial crisis, IBM released the theme report of *Smarter Planet: the Next Generation of Leaders Agenda* in New York, putting forward the concept of "Smarter Planet" to deal with the crisis, that is, using new generation of information technology in all walks of life. In February 2009, IBM put forward the "Smart City Breakthrough in China" strategy in Beijing, and signed a smart city co-construction agreement with more than 10 provinces and cities in China successively. Since then, the "smart city" concept was accepted gradually. The development of intelligent city was considered to help promoting the sustainable development of economy, society and environment, resource coordination in the city, alleviating the "big city disease", improving the quality of urbanization, so it quickly became an important fulcrum to meet the challenges of the city in the world.

At present, the planning and construction momentum of the smart city is like a single spark which can start a prairie fire. In the Americas, the IBM of United States first proposed the "Smarter Planet" program and carried out the smart city practice with the support of the government; in Europe, *Europe 2020* proposed to build a smart, sustainable, inclusive Europe, and put the smart city as an important part of it, hoping through the smart city construction, to release the potential of current infrastructure and capital to the greatest extent, and through research and development of new products and services to meet social and environmental challenges. In the Asia-Pacific region, South Korea and Japan launched u-Korea and u-Japan national strategic plan successively as early as in 2004, Singapore proposed to build "smart island" plan in 2015, and China also actively promoted the smart city construction. Till September 2013, China had a total of 311 cities under construction or planning to build smart city. In 2013, the Ministry of Housing and Urban-Rural Development of the People's Republic of China announced two batches of national smart city pilot lists, including a total of nearly 200 pilot cities (districts and towns). This series of initiatives reflected that China paid high attention to smart city construction and expressed its positive development attitude.

2 A Review on Intelligent City Development in Europe

The European Commission has been actively promoting the construction of smart city, and took the smart city as an important means to promote sustainable development, hoping through the smart city construction, to maximize releasing the potential of current infrastructure and capital, and through research and development of new products and services to meet social and environmental challenges.

In July 2005, the EU proposed the "i2010" strategy (Atlantic Council 2013), which was aimed to promote the EU members to develop communication technology and build new network infrastructure, innovation media and so on, and in 2010, to develop three key areas: the elimination of barriers in the internal markets, the establishment of a unified European information space network, increasing the scientific research investment in the field of information communication technology, and encouraging enterprises to apply new communication technology, thereby improving labor productivity, popularizing cultural education, and popularizing computer and the Internet and other related knowledge in the Member States, to enhance the overall knowledge level of the public.

In November 2006, the Living Lab project was launched, adopting the collaborative innovation method to gather the wisdom of various countries. Its core value is the improving insight on R&D and improving the driving force to transform new applications and solutions of scientific and technological achievements into real-world. Each country's independent living lab can be interconnected through the European Network of Living Labs (ENOLL) to mobilize the wisdom of collective innovation.

In January 2007, the EU 7th Framework Program was launched, with a duration of 2007–2013. This is the largest science and technology cooperation project in the world, with a total budget of 50.521 billion Euros. The main content of its research is focused on the international frontier and competitive scientific and technological difficulties, with the features of high level research, vast involved areas, strong investment, and many participating countries. Among them, the projects related to intelligent city include Radio Frequency Identification (RFID) technology and Internet of Things technology.

In March 2009, the European Commission put forward the *Information Communication Technology R&D and Innovation Strategy*, calling for increased support and investment in information technology R&D and innovation, making the EU leading the world in the field.

In 2009 "European Initiation Smart Cities", it presented the "Indicative Roadmap" for the development of smart city, which focused on strategic objectives, specific objectives, actions taken to achieve the goals, public and private investment and key performance indicators.

In December 2009, the European Commission published the "Internet of Things-Action plan for Europe[1]", which put forward 14 action contents about 9 aspects, including Internet of Things management, security assurance, standardization, research and development, openness and innovation, common agreements, international dialogue, pollution management and future development and so on. In addition, the program also described the application prospects of the EU Internet of Things technology, and put forward 10 policy recommendations about improving the Internet of Things management of the government, and promoting the development of the Internet of Things industry in the EU.

In March 2010, the European Commission introduced the *Europe 2020* to enable the EU to achieve smart growth, sustainable growth and inclusive growth. The "European Digital Agenda" was one of the flagship programs to promote economic growth, with the aim of achieving stable, sustainable and comprehensive economic growth through the good application and wide adoption of information communication technology and proposed seven key areas: first, to establish a single dynamic digital market in EU; second, to improve the development of information communication technology standards to improve the operability; third, to enhance network security; four, to achieve high-speed and ultra-high-speed Internet connection; five, to promote the research and innovation of the leading field of information communication technology; six, to improve the digital attainment, digital skills and digital inclusion; seven, to use the information communication technology to generate social benefits, such as use the information technology for

---

[1]Refer to http://www.eesc.europa.eu/?i=portal.en.ten-opinions.18007 for details.

energy conservation and environmental protection, and helping the elderly and so on.

In June 2011, the EU Energy Commission announced the EU's New Smart Cities and Communities Initiative report, which indicated from the perspective of energy development that it was the best time to build a smart city, and the city needed to use smart method to solve the problem, the industry needed to use smart method to improve efficiency, and residents needed to use smart method to get clean, safe and cheap energy.

In 2011, the European Innovation Partnership for Smart Cities and Communities was launched by the European Smart Cities Innovation Partnership project. The project was divided into two components: the High Level Group and the Smart Cities Stakeholder Platform. The project aimed to apply energy, transportation and information communication technology in a number of pilot cities through the establishment of partnerships between enterprises and cities. In 2012, the project received 81 million Euros in EU funding support, including two directions of transportation and energy, and was implemented as a pilot "Lighthouse Project".

In 2013, the total investment of the European Smart Cities Innovation Partnership Project reached 36.5 million Euros, and Information Communication technology (ICT) was classified as new project, which consisted of three research areas together with previous energy and transportation. At the same year, the project transferred into the framework of "Horizon 2020" for operation.

At the same time as the EU government promoted the construction of intelligent city, the research organizations, enterprises also launched corresponding exploration and set up a project cooperation and promotion platform (see Table 1).

2.1   UK

The Gloucester Smart House project was launched in 2007, with a house as a pilot, installing sensors inside the room and aggregating information to the central computer. The purpose is to intelligently assist the elderly living alone at home, such as controlling the temperature of the room according to the location of the person, monitoring the heart rate and blood pressure of the elderly, and giving the alarm timely in case of danger and sending the information to the guardian. The project ended in March 2008 with 40 lectures and 8 papers published in total.

The Civic Dashboard of Birmingham[2] is rich in content, and one of the most important features is to mark the different topics reflected by residents in the administrative region on the map by the geographical location, and to classify them, displaying real-time attention degree to similar issues to help the government's set priority in decision-making. Meanwhile, this site also contains weather forecasts

---

[2]Source of image: http://civicdashboard.org.uk.

Table 1 Table of EU intelligence/smart research institutions

Research institutions	Introduction
City Protocol[a]	A consortium consists of cities, businesses, nonprofit organizations, universities, and research institutions, which aims to assist urban transformation by formulating city protocol, and to promote the construction of smart city by partnering with other important Internet organizations
CitySDK[b]	The developer program of EU organization hopes to connect data across unifying the application interfaces of cities to provide developers with a range of modular tools
Citymart[c]	It is committed to linking city researchers and city managers around the world to create a platform for solutions of city problems, building bridges between research and demand, and more than 50 cities have participated in this project currently
The Climate Group[d]	An independent nonprofit organization that works with departments of companies, cities and nations, which is committed to lead the "green revolution" and build a low-carbon, intelligent, and bright future for the people all around the world
Smart City Expo[e]	An Smart City Expo held in Barcelona every year, which scholars from the field of intelligent city research are invited to give keynote speeches, and the leaders of the cities are invited to introduce the city practice, and the participants include the researchers of smart city, the companies which provide the technologies, and the mayors who desire to develop smart city
Metropolis[f]	A metropolitan research association composed of representatives from more than 120 cities around the world to discuss the problems faced by metropolitan areas and megalopolis and its solutions in the form of international forums
New Cities Foundation[g]	A Swiss nonprofit organization dedicated to improving the quality of life and work of the cities around the world in the 21st century, with particular attention to emerging cities in Asia, the Middle East, Latin America and Africa, building a unique platform between the public, government and academic institutions
Smart City Council[h]	The Smart City Council is an important promotion organization for smart city, aiming at accelerating the cities' process towards the smart and sustainable city. It publishes the Smart City Handbook, provides concise information for city leaders, diagnoses the city's technology issues, and portray the technical roadmap for building smart cities. Its think tank includes industry leading technology companies, research institutes, universities and so on
WeGO[i]	An international organization which is aimed at practicing and promoting the construction of intelligent government worldwide, and achieving green development goals through information communication technology, thus improving the efficiency and transparency of government management, and enhancing the publics' quality of life

[a]Refer to http://www.cityprotocol.org for details
[b]Refer to http://www.citysdk.eu for details
[c]Refer to http://www.citymart.com for details
[d]Refer to http://www.theclimategroup.org for details
[e]Refer to http://www.smartcityexpo.com for details
[f]Refer to http://www.metropolis.org for details
[g]Refer to http://www.newcitiesfoundation.org for details
[h]Refer to http://www.smartcitiescouncil.com for details
[i]Refer to http://www.we-gov.org for details

Fig. 1 Intelligent Trash Cans in London (*Source of image* http://www.inewidea.com/2012/02/03/44574.html)

and other conventional convenience services, which deserves to be called the true state of the city portrayal.

BedZED is a low-carbon sustainable development community in London, which was completed in 2002. During the construction, it used environment-friendly materials, by means of intelligence to achieve thermal isolation, intelligent heating, natural lighting and other design, and integrated solar energy, wind energy, biomass energy and other renewable energy, to achieve zero emissions of energy supply system, recycling of water-saving systems, and zero energy consumption of the heating system.

During the 2012 Olympic Games, there were a number of intelligent trash cans installed in London, which added LED electronic display on both sides of the traditional trash cans, rolling broadcast news. At the same time, the trash cans were also wireless hotspots, providing free Wi-Fi for the surrounding people (see Fig. 1). There were sensors inside the trash cans, which could automatically close the trash cans when the they were full and notify the sanitation department to clean up.

2.2 The Netherlands

Amsterdam, the capital of the Netherlands can be described as a model of intelligent city construction in Europe, which is also the earliest city starting intelligent city

Fig. 2 The spatial distribution of smart city construction project in Amsterdam (*Source of image* http://AmsterdamOpent.nl)

construction in the world. Amsterdam has launched a number of smart city practice projects

(See Fig. 2), mainly in the following five aspects.

1. Living

Amsterdam is the second largest city in the Netherlands, with more than 400,000 families. The annual carbon dioxide emissions of all households accounts for one-third of total in Amsterdam. Through the application of smart and energy-saving technology, carbon emissions are greatly reduced. The following practical projects can test which technical means are more effective (including from the renovation of the canal building to the installation of smart meters):

- Almere Smart Society
- Energy Management Haarlem
- Geuzenveld: Sustainable Neighborhood
- Groene Grachten
- IJburg: Fiber-to-the-Home
- IJburg: YOU decide!
- Smart Challenge
- West Orange

2. Working

Amsterdam brings together companies of all sizes and types, from small shops to multinational companies, the old houses beside the canal to the metal-and-glass office buildings. Energy-saving technologies can reduce energy consumption at work, such as working at the "Smart Work Center" instead of wasting time in traffic jams. The practical projects of the work are as follows:

- Distributed production: Fuel Cell Technology
- IJburg: Smart Work@IJburg
- ITO

- Monumental Buildings
- Municipal Buildings
- TPEX: Smart Airmiles

3. Mobility

The mobile vehicles in Amsterdam include cars, buses, trucks and cruise ships and so on, with carbon dioxide emissions accounting for one-third of the entire emissions in Amsterdam. The strategy focuses on the sustainability of transportation, such as the introduction of new logistics concepts, dynamic traffic management and even electric bicycle charging devices throughout the city. The practical projects of the transportation are as follows:

- E-Harbours: ReloadIT
- Moet je Watt: Charging System
- Ship to Grid
- WEGO Car Sharing

4. Public facilities

In order to achieve the zero impact of climate in 2015, the Government of Amsterdam has played a pivotal role in sustainable public spaces, including public buildings and public transport facilities. Projects in the field of public facilities focus on the application of smart solutions in schools, hospitals, stadiums, libraries and streets, as follows:

- Climate Street
- E-Harbours: Innovative Energy Contract Zaanstad
- Health-Lab
- Nieuw West: Smart Grid
- Smart Schools Contest
- Smart Sports Parks
- Swimming Pools
- Zuidoost: Laws and Regulations
- Zuidoost: Energetic Zuidoost
- Zuidoost: Stakeholders in the Driver's Seat

5. Open data

Like many metropolises in the world, Amsterdam also has its own government open data platform. Data supports the information society, and open data can effectively support decision-making, so this is one of the important aspects of smart city construction in Amsterdam. The practical projects of open data are as follows:

- AmsterdamOpent.nl
- Apps for Amsterdam

2.3   Sweden

As the capital of Sweden, Stockholm is facing the same traffic congestion as other metropolises. The Stockholm government uses intelligent means to smooth traffic. It launched the "Smart Transportation" project in 2006, and worked with IBM to develop a complete vehicle automatic payment system to limit traffic flow by imposing a "vehicle congestion tax" on vehicles entering the urban area. The system adopts advanced vehicle identification technology, and takes advantage of image enhancement and front and back license plate comparison technology to analyze the entire image and search for preset mode. The algorithm simulates the function of the human eyes, constantly moving the image until it finds the best viewing angle and identifies the desired pattern, thereby restoring the license plate that is usually unrecognizable. Relying on this technology, it can capture the time for vehicle entering and leaving the toll area, and automatically complete the charges.

The use of smart transportation system reduced traffic congestion by 25% in Stockholm, 50% for traffic queuing, 8–14% for road traffic emissions, and 40% for greenhouse gas emissions such as carbon dioxide. For its excellent performance in environmental protection, Stockholm was rated as the annual smart city by the Intelligent Community Forum in 2009, and was rated as the first "European Green Capital" by the European Commission in 2010.

2.4   Denmark

The Copenhagen in Denmark is known as the city of bicycles, so the practice of smart city in Copenhagen is concentrated in the field of bicycles. The introduced smart bicycles have a simple electric charge device that accumulates energy during braking and going downhill, and releases when accelerating and going uphill, saving effort for the rider (see Fig. 3). Public bicycles are important means of transport for the government to promote green travel (see Fig. 4). The new public bicycles have screens that replace the traditional coinage charge. Renting becomes electronic, the public complete the rental and payment through the mobile terminal, and the location of each bicycle can be monitored in real-time to understand the distribution amount of the public bicycle in different partitions.

2.5   Spain

In 2008, Barcelona launched the 22@Urban Lab project to obtain information about the interaction of these technologies practices with space by experimenting with a wide range of services and technologies in public spaces. A total of 43 project plans were received, of which 14 had been implemented. For example, the

Fig. 3 The smart bicycles in Copenhagen (*Source of image* http://lucaslaursen.com/ copenhagen-pioneers-smart-electricbike-sharing)

Fig. 4 The public bicycles in Copenhagen (*Source of image* http://senseable.mit.edu/ copenhagenwheel)

"Street Lighting" project was designed by Eco Digital to install 12 intelligent streetlights with temperature, humidity, sound and pollution sensors as well as used as Wi-Fi hotspots.

The "Meter Reader" project was driven by the Municipal Housing Authority, which installed automatic reading devices of gas, electricity and water meter for 150 construction units of 3 buildings; in the "Traffic Control Cameras" project, in addition to the installed traffic surveillance cameras, it will also install an intelligent system in the future, so that it can adjust the time length of the green lights based on the number of motor vehicles on the street to reduce the times of traffic lights conversion.

In 2011, the Barcelona government set up a public-private partnership platform which was LIVE (Logistics for the Implementation of the Electric Vehicle) Barcelona, in order to promote the widespread use of electric vehicles. At present, Barcelona already has 240 electric car charging stations, most of which are operated by grid operator Endesa Cepsa.

In 2012, the BCN Apps Jam for DEMOCRACY project which was launched by the Apps4bcn website, encouraged developers to develop related mobile applications for Barcelona, including electronic voting systems, electronic petitions, and polls, etc.

In 2012, the "Smart Service Delivery Platform" project was dedicated to the installation of a large number of sensors in urban spaces to facilitate access to data that was difficult to accurately measure in the past, such as the sensors measuring the amount of solid waste, monitoring whether the parking lot is occupied, and the sensors measuring air pollution and noise and so on. All of the information will be integrated into a unified information platform to provide a more complete and convenient information service for the public and the government.

The "Smart City Expo" was held in Barcelona for two consecutive years in 2013 and 2014, with 51 cities and more than 6000 people and hundreds of enterprises participating. The Expo established a communication bridge between the smart city technology companies and government, to spread the advanced construction experience of smart city to participants.

2.6   France

In 2011, Paris organized a design competition, with the theme of "Smart Street Furniture", to transform the streets facilities such as street lamps, trash cans, bus stations and billboards with intelligent means, in which designers and enterprises participated and finally 40 designs selected were implemented.

Concept Shelters is a digital bus station with a large touch screen display, through this display, the public can easily communicate with the relevant government departments and access the latest cultural activities (see Fig. 5).

At the same time, the station also provides free mobile phone charging interface for public emergency charging. Most of the electric energy required for the station comes from the solar photovoltaic panels of the roof. In this sense, it is also a "green" bus station.

NAutreville is a translucent touch screen that uses augmented reality technology, similar to a camera using the same technology (see Fig. 6). This touch screen can be rotated 360°, while the cultural activities, attractions information of the

Fig. 5 Digital bus station in Paris (*Source of image* http://www.bustler.net/index.php/article/high_tech_bus_stop_by_patrick_jouin)

Fig. 6 Translucent touch screen in Paris (*Source of image* http://ooh-tv.fr/2012/04/03/mui-paris-nautreville-une-fentre-enrichie-sur-la-ville)

neighborhood can be overlaid to the real images, in addition, it can also provide road navigation services for tourists. At present, there are a number of NAutreville put into use in famous attractions and community parks in Paris.

3   An Overview on Intelligent City Development in America

3.1   The US

In 2008, IBM put forward the concept of Smarter Planet, which has been recognized by the US government, and the smart city became one of the important strategies to deal with the economic crisis. In December 2012, the *Global Trends 2030*, published by the National Intelligence Council, noted that the four most influential technologies for global economic development were information technology, automation and manufacturing technology, resource technology, and health technology, in which "smart city" was part of the information technology content. The US government promotes the construction of smart city through the financial funds, and guides enterprises in related fields into scientific research, and encourages innovation.

Information infrastructure, intelligent grid, intelligent transportation, smart medical and other construction are the key points of current smart city construction in US.

In November 2008, Sam Palmisano, CEO of IBM, presented the concept of Smarter Planet in a speech at the Council on Foreign Relations, that is, fully applying the new generation of information technology to all walks of life.

In September 2009, IBM cooperated with Dubuque Iowa in the Midwest of United States to build the first "smart city", which adopted information technology to implement the city's digital system, the establishment of three systems of smart water, smart energy, smart transportation, and to enhance the efficiency of urban operation through the integration of information. The experimental team has established water monitoring system to monitor, analyze and warn in real time, and to detect leaks. After 15 weeks of experiments, household average water consumption fell by 6.6%. After installing the intelligent energy monitoring device, residents could view the household power consumption, and stagger electricity peak, and thus the average household power consumption was reduced by 11%.

In February 2009, the United States released the "Economic Recovery Plan", intending to invest 11 billion US dollars to build a new generation of smart grid where a variety of control equipment could be installed. In June the same year, the United States Department of Commerce and the Department of Energy jointly issued the industry standards for the first batch of smart grid, which marked the US smart grid project officially being launched.

In April 2009, San Jose launched the Intelligent Road Lighting project, whose network control technology was without the constraint of lamps, effectively saving energy, reducing operating costs, implementing remote monitoring, improving the quality of service and providing other benefits for a variety of outdoor and indoor lighting markets. Based on the efficiency of new lamps, through the early troubleshooting for failure street lamps, power failure detection and light output balance and dimming and other functions, the intelligent control networking technology can reduce costs and improve services, while making the city streets, roads and highways more safe and beautiful.

In March 2010, the Federal Communications Commission (FCC) officially announced the next 10 years development plan of the United States high-speed broadband network, which would be achieved through market incentives, resource security, universal service and application promotion and other efforts to increase the broadband speed by 25 times, that is, by 2020, to make the average speed of Internet transmission for 100 million US households improved from 4 Mbps in 2010 to 100 Mbps. In 2011, the FCC passed the previously announced reform plan for the universal service fund and the inter-operator compensation system and began implementing the fund promotion programs which annual budget up to $4.5 billion.

3.2   Brazil

Rio de Janeiro is an important city in Brazil, which hosted the 2014 World Cup and the 2016 Olympic Games, making it a hotspot city with the attention of the world. The landslides that occurred in 2010 caused a large number of casualties, and similar natural disasters also accelerated the pace of government's efforts to promote smart city construction.

Fig. 7 Center of operations in Rio de Janeiro (*Source of image* http://thenextweb.com/la/2011/07/13/how-data-is-making-rio-de-janeiro-a-smarter-city)

In 2010, the Rio de Janeiro government worked with IBM and Oracle to set up the Center of Operations to collect and centrally process information in the city (see Fig. 7), such as to position the garbage truck location through GPS, and to set the best action line for the city running more efficiently.

Another important measure is data disclosure. At the beginning of the establishment of the Center of Operations, Rio de Janeiro had set the transparence as an important principle. The city-related data and information was opened to the public, all the media could visit the Center of Operations and accessed to relevant information, and the public could also visit the database through the website. This openness maximized the use of smart city data.

4 An Overview on Intelligent City Development in Asia-Pacific Region

4.1 Japan

In 2004, Japan launched the "u-Japan"[3] strategy, hoping through the national information strategy to promote a new round of scientific and technological revolution, popularize the digital to all aspects in life, address the social problems such as population aging, and enhance the overall competition force of the country.

---

[3] 'u' refers to 'ubiquitous', which means 'immanent, omnipresent'.

In 2009, Japan also proposed the "i-Japan" strategy, focusing on e-government governance, health care information services, education and personnel training and other three major public utilities, aimed at achieving the digital society of people-oriented, peace of mind and full of vitality in 2015. Compared with the universalness of the "u-Japan" strategy, this strategy is more concerned with the ease of use of information technology to break through barriers to digital technology to ensure information security, and ultimately to permeate through the digital and information technology to the economic society.

Clean energy strategy is the main energy strategy promoted by the Japanese government. The "Feed-in Tariff" project launched in June 2012 has driven the development of the photovoltaic solar power industry, and the smart grid was necessarily developed simultaneously. Solar energy cannot be completely consumed during the day and need to be stored at night in battery storage, therefore, batteries and fuel cells become key components of the intelligent community, helping to balance the grid. The Advanced Metering Infrastructure and Energy Management Systems allow users to understand the details of energy consumption in real time, to avoid the peak demand of electricity and to charge the electric cars at a lower cost.

The Ministry of Economy, Trade and Industry (METI) is an important department of intelligent city construction. The local government can propose intelligent city solutions jointly with consulting firm. If the proposal is adopted by METI, it will be supported by the country's fiscal. The local government is the bridge between the enterprise and the country. In 2010, the first round of scheme solicitation started. 20 candidate cities submitted master plans, of which four cities passed the assessment and stepped into the implementation phase. In 2012, one year after the accident of Fukushima Nuclear Power Plant occurred, the topic of second-round scheme solicitation was city recovery and economic reconstruction after hard hit.

4.2   Korea

In 2004, South Korea put forward the "u-Korea" strategic plan, hoping to promote economic development and social reform in Korea through digital system, network, visualization, and intelligence. According to the overall policy plan confirmed in March 2006, the development period of u-Korea is 2006–2010, and the maturity period is 2011–2015. Through the construction of ubiquitous information perception network, providing convenient living services for the public, and enhancing the convenience of the publics' living, it can also promote the development of information technology-related industries, and strengthen the country's competitiveness.

u-Korea contains five key construction fields: ① Friendly Government, which provides smart administrative office management, mobile public service and smart voting system; ② Intelligent Land, which builds intelligent traffic management network, and electronic passport entry management system; ③ Regenerative Economy, which uses new business marketing model to promote online payment; ④ Tailored u-life Service, which provides electronic identity card, and creates a smart family life; ⑤ Secure Safe Social Environment, which establishes smart emergency response network, food and drug origin tracking system, and unmanned security system.

In 2009, South Korea adopted the "u-City" comprehensive plan, which incorporated the u-city construction into the national budget, and vigorously supported the core technology localization, marking that the smart city construction had risen to the national strategic level. U-City is the detailed implementation of u-Korea, on the one hand hoping to make the public feel the convenience which smart technology provides for the life through the construction of smart city model, on the other hand to promote the development of related industries to make progress for the national economy.

U-City construction has developed two goals and four promotion strategies. The two goals are: ① to let u-City become a new engine of economic growth in South Korea, and nurture u-City new industry; ② to promote u-City construction mode abroad. The four promotion strategies are: ① to build the system platform; ② to develop core technologies; ③ to support u-City industry development; ④ to cultivate talent.

The "New Songdo City", which began construction in August 2009, is about 65 km west of Seoul and has a reclamation area of about 607 ha, jointly developed by Gale International and POSCO E&C. The communities, hospitals and companies in New Songdo City share information all-around in real-time, and public housing, streets and office buildings connect together through the network. The public can enjoy education, health care, shopping and a series of life services in the city with the smart card.

4.3   Singapore

In 1992, Singapore proposed the Smart Island Plan (IT2000), which planned to build a nationwide high-speed broadband multimedia network in ten years, popularize information technology, and establish a more closely linked electronic community at the regional and global levels, building Singapore into a smart island and Global IT Center.

In 2000, Singapore proposed the "21st Century Information Communications Program" (Infocomm21), which promoted Singapore to a global information communications capital market based on the Smart Island. Singapore planned to be a leader in all areas of the Asia-Pacific region, including information communications industry research and development, venture capital, intellectual property, education and new ideas, and thus to be a global leader in information communications application.

The program set six goals: ① Business Online, to develop Singapore to a global e-business center, leading the world's B2B or B2C e-commerce; ② Government Online, the Singapore government was committed to becoming the world's premier e-government to provide better services for people; ③ Singaporeans Online, to become an e-life society, the capabilities of its national application of information communications will lead the world; ④ Infocomm Talent Capital, to become the human resources center of information communications and the hub of e-learning, and to shape the environment which will attract high-quality information communications talents to develop and provide excellent field for electronic learning; ⑤ Pro-Business and Pro-Consumer Environment, to strengthen the formulation and revision of policies, laws and regulations, to develop an information communications industry conducive to enterprise development, and to enable consumers glad to use the environment of information communications services to promote the development and growth of new economies; ⑥ Infocomm Hub, to connect with the world's leading ICT centers, R&D centers and e-commerce markets, and to make ICT becoming the main driver of Singapore's e-economic growth.

The IN2015 program announced in 2006 was the action hosted by the Infocomm Development Authority (IDA), and was participated jointly by government and enterprises, which planned to build Singapore into a global leader in informatization within 10 years. It planned, by 2015, to realize the additional value of information industry reaching 26 billion US dollars, the export output value of information industry reaching 60 billion US dollars, and adding 80,000 new jobs, and to achieve more than 90% of home broadband coverage, the home computer owning rate of school-age children reaching 100%.

4.4 Malaysia

In 1995, Malaysia proposed the Multimedia Super Corridor (MSC) program, covering the Kuala Lumpur City Center, the Putrajaya Government Administration Center, Cyberjaya, the Hi-Tech Incubation Innovation Park and Kuala Lumpur International Airport. The MSC program included seven "flagship programs": e-government, smart school, telemedicine, multi-purpose smart card, research and development center, borderless marketing center and global manufacturing network. The entire MSC program will last to 2020.

The Cyberjaya, which is 40 km from the Kuala Lumpur, covers an area of 2800 ha. It is the core project which can best reflect the "smart city", and is known as "Oriental Silicon Valley". The whole Cyberjaya is divided into software development zone, system integrated zone, telecommunications and network service zone, animation and film zone, electronic zone, education and training zone and so on. The Malaysian government plans to build the Cyberjaya into world's chip production center by 2020, and to develop multimedia products, applying multimedia to the fields such as education, market development, health and medical research.

Reference

Atlantic Council (2013) Envisioning 2030: US strategy for the coming technology revolution. Brent Scowcroft Center on International Security, Washington, D.C.

Chapter 2
Status of Intelligent City Construction in China

Abstract To accommodate and resolve the problems associated with rapid urbanization, the Government of China has attached great importance to the development of intelligent city. This chapter provides a comprehensive introduction of intelligent city construction in China, including the smart city initiatives in China since 2012 and the institutions involved ranging from national level to local level, the characteristics and categories of intelligent city construction modes in China, as well as the technological requirements and critical technologies in building intelligent city system. Overall, the Intelligent city construction in China is growing vigorously and the construction of many cities involves many concepts such as Internet of Things, ubiquitous network and cloud platform, which have been interpreted by experts from various angles. However, it is undeniable that most people still lack deep understanding about how these terms should be applied in specific construction, so eventually, all constructions tend to be similar and lack breakthrough. Given that China intelligent city construction is still at primary stage with many problems and challenges to be overcome, intelligent city construction in China is suggested to be promoted in three steps gradually at the end of this chapter.

Keywords Intelligent city · China · Policy

1 A Review of Intelligent City Construction in China

1.1 The Genesis of Intelligent City Construction in China

The report to 18th National Congress of the Communist Party of China put forward that in the national strategy of China in the next period, we must adhere to the route of "four modernizations synchronization", that is, "adhere to the new industrialization, informatization, urbanization and agricultural modernization with Chinese characteristics, to promote the deep integration of informatization and industrialization, the virtuous interaction of industrialization and urbanization, the coordination of urbanization and agricultural modernization, and to promote the

© Springer Nature Singapore Pte Ltd. and Zhejiang University Press 2018   19
Z. Wu, *Intelligent City Evaluation System*, Strategic Research
on Construction and Promotion of China's Intelligent Cities,
https://doi.org/10.1007/978-981-10-5939-1_2

simultaneous development of industrialization, informatization, urbanization and agricultural modernization". Urbanization is the necessary means for China to realize economic growth, stimulate domestic demand and arrange overall urban and rural development, and so on. And when confronting the problems of population, economy and environment in urban development, how to realize the sustainable development of urbanization and realize the residents' life, economic achievements, urban construction and ecological carrying capacity well matched, will be the important issue.

Therefore, the "intelligent city" concept which takes "high-tech as the core, intelligent technology as the means" came into being. By extracting the elements of the city, realizing the real-time monitoring of these elements by means of technical, which reflects the influence and function of the elements of the city entity at the virtual level, so as to re-sort, integrate and associate these relations, furthermore, the results are reflected in the construction of the entity city, playing a guiding role to the development of urbanization. As a new urban development model, "Intelligent City" will reverse the status of lack of systemic urban construction and urban construction fragmented due to poor information for the systems do things respectively in their own way, gradually healing all kinds of "urban disease" to realize Intelligent Urbanization in China.

1.2   Status of Intelligent City Construction in China

China attaches great importance to the development of intelligent city. In order to standardize and promote the healthy development of the intelligent city, the Office of Ministry of Housing and Urban-Rural Development officially released *Notice on Carrying out the National Smart City Pilot Work* on November 22, 2012, and issued two documents of *Temporary Management Approach of National Smart City Pilot* and *National Smart City (District/Town) Pilot Indicator System (Trial)*, which began pilot city application at once. The documents pointed out that intelligent city construction is an important measure to implement innovation-driven development, promote new urbanization and build a well-off society in an all-round way of the Party Central Committee and the State Council; in the process of construction, the housing and urban and rural construction departments which are responsible should strengthen organization, coordination, supervision and assessment.

On Jan. 29, 2013, after the application of local cities, the first reviewing of the Provincial Ministry of Housing and Urban and Rural Construction Competent Departments, experts comprehensive reviewing and other procedures, the Ministry of Housing and Urban-Rural Development announced the first batch of national smart city pilot list, with a total of 90 pilot cities (see Table 1). In Aug. 2013, the Ministry of Housing and Urban-Rural Development announced the second batch of national smart city pilot list, confirming 103 pilot cities (see Table 2). In addition to the first batch of 90 national smart city pilot announced by the Ministry of Housing and Urban-Rural Development previously, by the end of 2014, the national smart

Table 1   The first batch of national smart city pilot list

Administrative level	Quantity	Pilot
Capital City	5	Wuhan, Zhengzhou, Shijiazhuang, Taiyuan, Lhasa
Prefecture City	29	Changzhou, Wuxi, Wenzhou, Zhuhai, Tongling, Weihai, Dongying, Huainan, Jinhua, Wuhu, Zhenjiang, Nanping, Taizhou, Zhuzhou, Dezhou, Pingxiang, Qinhuangdao, Ya'an, Wuzhong, Xianyang, Liupanshui, Tongren, Luohe, Hebi, Liaoyuan, Handan, Langfang, Changzhi, Wuhai
County City (County)	17	Xintai, Shouguang, Zhuji, Wanning, Shaoshan, Qian'an, Korla, Chengdu Pixian, Zhaodong, Daqing City Zhaoyuan County, Jiamusi City Huanan County, Jiyuan, Xinzheng, Changyi, Kuitun, Feicheng, Panshi
City District	14	Chaoyang District of Beijing, Dongcheng District of Beijing, Jinnan District of Tianjin, Shangcheng District of Hangzhou, Zhenhai District of Ningbo, Shunde District of Foshan, Panyu District of Guangzhou, Jiang'an District of Wuhan, Nan'an District of Chongqing, Wudang District of Guiyang, Cangshan District of Fuzhou, Wuhua District of Kunming, Pinglu District of Shuozhou, Yuhui District of Bengbu
New District, Ecological Zone, Demonstration Area	22	Shanghai Pudong New District, Beijing Lize Financial Business District, Beijing Future Science Park, Sino-Singapore Tianjin Eco-city, Nanjing Hexi New Town (Jianye District), Guangzhou Luogang District, Suzhou Industrial Park, Chengdu Wenjiang District, Yancheng Chengnan New District, Shenzhen Pingshan New District, Zhuzhou Yunlong Demonstration Zone, Qinhuangdao Beidaihe New District, Nanchang Honggutan New Area, Jinan Xibuxincheng, Chongqing Liangjiang New District, Shenyang Hunnan New District, Dalian BEST City, Kunshan Huaqiao Economic Development Zone, Luoyang New Area, Yangling Agricultural High-tech Industries Demonstration Zone, Changsha Dahexi Pilot Zone, Pingtan Comprehensive Experimental Zone
Town	3	Zhangpu Town of Kunshan City, Lecong Town Shunde District of Foshan City, Bo Jia Town of Liuyang City
Total	90	

Table 2 The second batch of national smart city pilot list (103 in 2013)

Administrative level	Quantity	Pilot
District, City	83	Beijing Economic Technological Development Area, Tianjin Wuqing District, Tianjin Hexi District, Chongqing Yongchuan District, Chongqing Jiangbei District, Tangshan Caofeidian District, Yangquan City, Datong City, Jincheng City, Hulunbeier City, Ordos City, Baotou Shiguai District, Qiqihar City, Mudanjiang City, Anda City, Siping City, Yushu City, Changchun High-tech Industrial Development Zone, Yingkou City, Zhuanghe City, Dalian Puwan New District, Yantai City, Qufu City, Jining Rencheng District, Qingdao Laoshan District, Qingdao High-tech Industries Development Zone, Qingdao Sino-German United Group, Nantong City, Danyang City, Jiangsu Wuzhong Taihu New District, Suqian Yanghe River New Town, Kunshan City, Fuyang City, Huangshan City, Huaibei City, Hefei High-tech Industrial Development Zone, Ningguo Port Ecological Industrial Park, Hangzhou Gongshu District, Hangzhou Xiaoshan District, Ningbo District (including Haishu District, Meishan Bonded Port Area, Yinzhou District Xianxiang Town), Putian City, Quanzhou Taiwanese Investment Zone, Xinyu City, Zhangshu City, Gongqingcheng City, Xuchang City, Wugang City, Lingbao City, Huanggang City, Xianning City, Yichang City, Xiangyang City, Yueyang Yueyang Tower District, Zhaoqing Duanzhou District, Dongguan Dongcheng District, Zhongshan Cuiting New District, Nanning City, Liuzhou City (including Yufeng District), Guilin City, Guigang City, Mengzi City of Hani-Yi Autonomous Prefecture of Honghe, Mile City of Hani-Yi Autonomous Prefecture of Honghe, Guiyang City, Zunyi City (including Renhuai City, Meitan County), Bijie City, Kaili City, Lanzhou City, Jinchang City, Baiyin City, Longnan City, Dunhuang City, Mianyang City, Suining City, Chongzhou City, Nyingchi Area, Baoji City, Weinan City, Yan'an City, Yinchuan City, Shizuishan (including Dawukou District), Urumqi City, Karamay City, Yining City
County, Town	20	Beijing Fangshan District Changyang Town, Tangshan City Luannan County, Baoding City Boye County, Shuozhou City Huairen County, Baishan City Fusong County, Jilin City Chuanying District Soudengzhan Town, Weifang City Changle County, Pingdu City Mingcun County, Xuzhou City Fengxian, Lianyungang City Donghai County, Lu'an City Huoshan County, Ningbo City Ningmei County, Lin'an City Changhua Town, Shangrao City Wuyuan County, Changsha City Changsha County, Chenzhou City Yongxing County, Chenzhou City Jiahe County, Changde City Taoyuan County Zhangjiang Town, Liupanshui City Pan County, Yinchuan City Yongning County

(continued)

Table 2 (continued)

Administrative level	Quantity	Pilot
Total	103	
Pilot expanding scope in 2012	9	Changzhou City Xinbei District, Wuhan City Caidian District (pilot in 2012 including Jiang'an District), Shenyang City Shenhe District, Shenyang City Tiexi District, Shenyang City Shenbei New District, Nanjing City Gaochun District, Kirin Science and Technology Innovation Park (Eco-science and Technology City), Changsha Dahe West Pilot Area Yanghu Ecological New Town and Binjiang Business New Town, Foshan City Nanhai District

Table 3 Smart city double pilot list (20 in 2013)

Implementation period	Quantity	Pilot
3 years (2013–2015)	20	Nanjing, Wuxi, Yangzhou, Taiyuan, Yangquan, Dalian, Harbin, Daqing, Hefei, Qingdao, Jinan, Wuhan, Xiangyang, Shenzhen, Huizhou, Chengdu, Xi'an, Yan'an, Yangling Demonstration Zone, Karamay

city pilots have reached a total of 193. In addition, the Ministry of Science and Technology of the People's Republic of China and the Standardization Administration of the People's Republic of China jointly identified the national "smart city" technology and standard pilot cities (referred to as "smart city double pilot") in 2013, selecting a total of 20 pilot cities this time, with the implementation period of 3 years (2013–2015), including nine sub-provincial cities (see Table 3).

On August 27, 2014, with the approval of the State Council, the National Development and Reform Commission (referred to as NDRC), the Ministry of Industry and Information Technology (referred to as MIIT), the Ministry of Science and Technology (referred to as MST), the Ministry of Public Security, the Ministry of Finance and the Ministry of Land and Resources, the Ministry of Housing and Urban-Rural Development (referred to as MOHURD), and the Ministry of Transport jointly issued *Guidance for Facilitating the Healthy Development of Smart Cities*, requiring all regions, and the relevant departments implemented the tasks proposed by this guidance, ensuring smart city construction in a healthy and orderly manner. The guidance has suggested that by 2020, a number of distinctive smart cities shall be built, the role of gathering and radiation shall be greatly enhanced, the comprehensive competitive advantage shall be improved obviously, and remarkable achievements shall be achieved in ensuring and improving people's livelihood services, innovating social management and maintaining network security and so on.

1.3 Institutions Involved in Intelligent City Construction in China

1. National level

(1) National Development and Reform Commission (NDRC)

As a major coordination department, NDRC is primarily responsible for the overall coordination of the construction of intelligent cities and the promotion direction for the construction of the urbanization reform by intelligent means.

In 2014, NDRC and MIIT led to draw up the Instructive Opinions on Robust Development of Smart Cities in China (hereinafter referred to as Opinions).

The Opinions completely summarized key points and main targets of the construction of Smart cities in five aspects, namely scientification of city planning and design, equalization of public service, refinement of social management, intelligent infrastructure, and modernization of industry development. Meanwhile, it put forward the development guideline of "Rational Promotion and Top Priority", which mainly emphasized the utilization of information resources and further strengthened related regulations on information security.

The Opinions emphasized the work coordination mechanism on national level in particular and suggested that NDRC, MIIT and other relevant departments shall establish an inter-ministerial coordination mechanism to solve main problems in the construction of intelligent cities.

(2) Ministry of Housing and Urban-Rural Development (MOHURD)

MOHURD is one of core departments in charge of the construction of intelligent cities.

As one of important departments responsible for promoting the construction of new-style urbanization, MOHURD puts forward and puts into practice the concept of intelligent cities and takes specific measures in 7 aspects, including policies and systems, standards and norms, technical research, industry team, construction fund, model city as well as theoretical research and publicity. These measures are as follows: releasing the management measures of Smart cities for the first time; undertaking the formulation of standards for national and international intelligent cities, coordinating and introducing bank and social capital of about RMB 100 billion into this area; establishing an industry alliance of intelligent cities with about 100 members including colleges and universities, research institutes, enterprises and investment organizations, among which there are 42 listed companies which cover all nodes in the industry chain of intelligent cities; selecting 15 pilot cities in different regions and with various scales and styles in China to conduct exploring work of intelligent cities.

In the process of promoting the establishment of intelligent cites, MOHURD collects various social elements by setting up platforms and achieves free matching and resource optimization according to the market rules and win-win principles. By doing so, it avoids the previous practice that the government takes care of everything and helps intelligent cities return to market and society.

Apart from paving the road for the construction of intelligent cities, MOHURD also needs to take another responsibility: evaluation and supervision of construction in later period. In China, currently the construction of intelligent cities is still in the primary stage and is moving towards further development, and the government is active in the construction of intelligent cities, and the construction is in full swing in all regions. However, in the construction of intelligent cities, detouring is unavoidable. Some cities go in for grandiose projects, pursue the biggest and the fastest without projection and invest blindly, which leads to waste of resources and is against the original intention of the construction of intelligent cities. Ultimately, the goal for the construction of intelligent cities is to improve the operational efficiency of cities and serve the public. Therefore, it is necessary to strengthen supervision as well as construct and operates according to local conditions, so as to guarantee the results of the construction of intelligent cities. At the same time, adjusting construction routes in every city in time will help avoid losses caused by adopting similar construction means in all conditions.

(3) Ministry of Industry and Information Technology (MIIT)

MIIT takes the pilot construction of intelligent cities as its priority.

As a competent department in the field of industry and information technology, MIIT deploys the work related to the construction of intelligent cities by means of plan formulation, pilot demonstration as well as summarization and popularization. Its key focus in work is to propel the application of information technology and intelligence in all fields, optimize the construction of information infrastructure and focus on hot spots in the information industry, and create a good technical and industrial environment for the construction and model innovation of intelligent cities.

In recent years, MIIT has formulated over 10 plans related to intelligent cities including the Twelfth Five-year Development Plan in the Internet Industry, the Twelfth Five-year Development Plan in Communication Industry, the Twelfth Five-year Development Plan in Broadband Network Infrastructure, the Twelfth Five-year Development Plan in International Communication, the Twelfth Five-year Plan in Telecommunication Network Code, Internet Domain Name and IP Address Resource and so on, covering several industries and fields including the application of information technology, information security, electronic information, software, communications plan and e-government and e-commerce plan through the Internet of Things. The release of these plans will strongly support the development of intelligent cities. Judging from the current situation, the information infrastructure is crucial to the construction of intelligent cities. MIIT should promote the evolution of the next generation information technology research, industrialization and networking, explore and develop broadband network construction in multiple modes according to characteristics of different regions to stimulate the settlement of optical fibers in urban areas, accelerate the construction of the new generation broadband wireless mobile communication network and popularize mobile internet to realize complete coverage of wireless network and wireless local area network in cities as well as invest more in 3G network, speed up its construction, upgrade its quality and make an overall plan for its evolution from TD to LTE.

2. Local level

The construction of intelligent cities in local level covers many aspects including urban infrastructure, industry management in information communication, urban transportation, medical treatment and public health, education as well as community management service, and the infrastructure construction involves in several sections such as water, electricity and energy, etc. However, due to lack of experience in early construction, although the overall progress in the construction of intelligent cities is made by the government, every sector in a project has its own priority. Particularly, when some information resources sharing involve departmental benefits and privacy, it is hard for sectors to open information to each other and make overall arrangements for the application of information technology. As a result, self-contained systems that lack efficient correlation are established in cities, which lead to low resource use efficiency and problems such as repetition and isolation.

Information isolation is the largest handicap in the construction of intelligent cities. MOHURD has greater superiority in putting specific projects of urban planning and construction management into practice, while MIIT specializes in integrating the application of new generation information technology and managing information resources efficiently. The two departments both have their own focuses, but lack unified targets and construction means. The isolation among sectors and information resources is more obvious in specific construction on a local level, but the current construction modes of intelligent cities cannot provide enough constructive experience and solutions.

Thus, China needs an intelligent cities construction system which can coordinate all departments and realize information resources sharing to make overall arrangements for the construction of intelligent cities so as to promote the construction efficiency. Of course, the construction led by a single department will easily meet many problems, so we can consider about starting from the perfection of superstructure and give priority to the establishment of overall leading departments in higher level, which can issue relevant instructive opinions about the construction and organize departments to coordinate and exchange with each other to solve problems such as business collaboration and information sharing faced in the process of the construction of intelligent cities.

2   Characteristics of Intelligent City Construction in China

2.1   Connotation of Intelligent City Construction in China

1. City construction intelligence

   With the urbanization rate in China reaches over 50% for the first time in 2011, city construction comes to a turning point. It is expected that in 15 years, there will be another 200 million people rushing into cities. The change means our future development pattern will turn from "rural areas encircling the cities" to "city development driven by prosperity in cities". Correspondingly, the scale of urban construction will also be expanded to meet the need of population.

   The target of intelligent city construction in the West is to use intelligent measures to manage cities, while China is facing problems of urban construction caused by large immigrants in terms of accommodation, transportation and education in the short term. Therefore, how to find a correct development strategy in the construction of intelligent cities by means of intelligence will be an important topic in intelligent city researches.

   Many western cities, such as London, Paris, Chicago, Prague and Florence, have all withstood the test of time and passed on their unique features, which have become city name cards now. Meanwhile, challenges faced by our urban construction are that many things left by the previous generation have disappeared in urban expansion and that we need to figure out how to remain the uniqueness of cities in rapid urbanization to stand the trial of urban development. What we need to do is to diagnose urban problems by means of intelligence and help cities make correct development decisions.

2. City industry intelligence

   Since the establishment of People's republic of China, we have been striving to develop industry, which results in economic growth in most cities, and the growth still heavily depends on the secondary industry, even in Shanghai. So, industry will still be an important means to develop the real economy in China in the next decades. The development trend of current global economic market tells us that we should pursue higher levels to win in the market.

   China is often referred as the "manufacturing factory of the word", which on the one hand confirms our manufacturing industry and on the other hand reflects the current shortage of innovation in our industrial development characterized, lack of brand advantage and focus on processing. Switzerland is not rich in resources, but it can produce watches that are as expensive as cars with intelligent manufacture. In contrast, manufacture in China is more dependent on equipment, technology and design from other countries.

   Therefore, influenced by intelligence, our country should, first, raise technical level and use our own science and technology in the industrial market to demonstrate our own advantages; second, promote scientific and technological innovation and strategically develop new industries to combine design and production closely and propel industrial development by strategic new products; third, straighten the market network and use it to master market information more accurately market information and find target customers in order to turn factories into companies.

3. City infrastructure intelligence

Normal operation of cities is based on infrastructure, so how to feel cities depends on the inner "channels" that cover all around cities. The construction of intelligent cities depends on the construction of intelligent infrastructure, including roads and streets, fire protection, environmental protection as well as systems of water, electricity and energy.

In addition, data will be another powerful weapon in changing our urban construction.

So, we need to improve internet construction as soon as possible to promote the broadband network coverage rate. Supported by network construction, intelligent cities will be able to dig information from data so as to lead the construction and development of all walks.

4. Three aspects joint-movement of cities

Against our special development backdrop, the construction of intelligent cities caters to the new trends and tasks in five aspects, namely industrialization, application of information technology, urbanization, adoption of the market principles and internationalization, proposed by the 17th National Congress of the CPC. Developed countries have already finished urbanization and industrialization, so their next construction target is the application of intelligence and information technology. Thus, we can then understand that as for China, the construction of intelligent cities not only is an important means to promote the combination of the application of information technology, urbanization and industrialization, but also plays an important role in the process of the adoption of the market principles and internationalization.

2.2 Characteristics of General Construction of Intelligent Cities in China

1. Powerful promotion of national policies

The construction of intelligent cities is an important step in the strategic deployment to realize synchronous development in the four aspects, so several departments in China have made it as their key focus in work and all cities have also set it as their next development target. By May 2011, all tier-one cities have put forward their detailed plans for the construction of Smart cities, while 80% tier-two cities also put it on the agenda. By September 2012, in the planning documents of 47 cities above the deputy provincial level, 22 have put Smart city construction on the agenda, accounting for 46.8%. By January 2013, 320 cities have invested 300 billion yuan in the construction of Smart cities (Zhou 2013).

In 2010, MOST recognized Wuhan and Shenzhen as pilot cities of the national project of 863 Smart cities. MIIT approved Yangzhou, Changzhou as pilot model cities of Smart city construction respectively in 2011 and 2012.

In April 2012, the Chinese Academy of Engineering published five pilots "Chinese Smart Cities", namely Beijing, Hangzhou, Wuhan, Ningbo and Xi'an.

In August 2013, the State Council issued Some Suggestions to Expand Domestic Demand of Information Consumption, which clearly pointed out that we should accelerate the construction of Smart cities and make pilots in some qualified cities.

MIIT published the first list of 90 pilot Smart cities (districts, counties or villages) in China, including Dongcheng District in Beijing, Wuxi City in Jiangsu Province in January 2013, and published another 103 pilot cities (districts, counties or villages) in August.

In October 2013, guided and supported by MIIT, the Chinese Industry Alliance of Smart Cities was established in Beijing. The alliance was launched by over a hundred large and medium-sized enterprises and public and research institutions, including China Electronic Chamber of Commerce and China Aerospace Science and Industry Corporation. There are already 4 national level alliances related with Smart cities in China, namely the Chinese Industry Alliance of Smart Cities, the National Strategic Alliance of Industrial Technology Innovation of Smart Cities, Chinese Promotion Alliance of the Planning and Construction of Smart Cities and Chinese Working Alliance of the Development and Promotion of Smart Cities. In addition, portals and research institutions such as China Academy of Smart Science, China Smart City Network and Govmade Internet Smart City Research Center were established. Meanwhile, supporting mechanisms, trade organizations and standard-setting bodies under MIIT, MOHURD and NDRC were accelerating deployment of Smart cities with their own features.

In November 2013, International Organization for Standardization approved our resolution of establishing Smart City Research Group with relevant persons in the Academy of Electronic Industry Standardization under the MIIT as its conveners and secretaries. Besides, the US, France, South Korea, Japan, Canada, Netherland, German, Britain and Singapore all show enthusiasm in participating the research group work. The approval of the resolution was a breakthrough in our international standardization of Smart cities, playing an important role in establishing our leader status in international standardization of Smart cities and promoting harmonious development of international and domestic standards of Smart cities.

In December 2013, the Engineering Research Center of Digital Cities under MOHURD and Microsoft China jointly declared to found the Joint Laboratory of Smart City Technology Solutions in the Engineering Research Center of Digital Cities under MOHURD to forge an important technical support platform for the development of our future Smart cities.

On January 1st, 2014, the Fourth China Smart City Conference was hosted in Beijing with the theme that developing Smart cities was the only way to realize new urbanization and with its content involving three focuses, new urbanization, Smart cities and big data. The conference will further promote the construction of Smart cities in China (Zhou 2014).

Conferences and files related to intelligent/Smart cities under all ministries and commissions between 2013 and 2014 (Chen 2014) are as shown in Table 4.

2. Rapid development of technical support

On the one hand, we shall accelerate network upgrading and raise the coverage rate. Main cities in China have already been covered with 3G network. Primary regions and superstores have also been covered by free Wi-Fi. Meanwhile, Households installed with optical fibers are growing. Broadband penetration and access bandwidth are increasing as well. By 2013, telephone subscribers reached 1.39 billion, among which over 1.1 billion were mobile users with 3G Network, accounting for 20%. The proportion of users with products with 4 M or above bandwidth is over 63% and new households covered by optical fibers exceed 43 million.

On the other hand, the application of information is expanding widely and the technology is increasingly mature. Application of information technology is expanding to various aspects of life including transportation, healthcare, consumption and tourism. Take Wuxi as an example, the city put forward the project of "citizen card" in 2009. In 2013, the card has accumulated such functions as transportation, micropayment and social security, especially, the card can help citizens trace food sources. In the future, the card can gradually realize more than 100 functions including borrowing books, renting bicycles and checking clinical cases.

3. Different construction situations in various regions

Judging from the regions of intelligent city construction, although many cities have already set up the construction target of intelligent cities, it is undeniable that cities along the east coast have strength in numbers, while cities in central and western regions lag (Yang 2012). In terms of the quality of intelligent construction, different cities also have different situations. After the process of statistics and quantification of the official, third-party and open data of 96 cities, the First Evaluation Report on the Development Level of Chinese Smart Cities has shown that in terms of smart city construction, 17 cities are in the stage of planning and design, 26 are in the stage of launch, 46 construction and promotion and no city is beginning to take shape. Thus, the development of different cities is uneven.

Judging from the aspects of intelligent city construction, developed cities in the east are paying more attention to application construction and focus on innovation management of cities and protection of people's livelihood, while developing cities in the central and -middle give higher priority to infrastructure (including information infrastructure), transformation of economy industry and perfection of social management (Sun 2014).

Table 4 Conferences and files related to intelligent/Smart cities under all ministries and commissions between 2013 and 2014

Date	Conferences/Files	Related organizations	Comments
January 29th 2013	National conference on the construction of pilot smart cities	MOHURD	Published the first list of 90 pilot Smart cities (districts, counties or villages)
February 5th 2013	Instructive opinions on robust development of internet of things	State Council	Issued and organized ten special action plans for the development of internet of things
May 3rd 2013	Notice of the application for national pilot smart cities in 2013	MOHURD	Published the second list of 103 pilot Smart cities (districts, counties or villages) on August 5th
August 8th 2013	Some suggestions to expand domestic demand of information consumption	State Council	Formally pointed out to develop pilot and model smart cities in some qualified cities
November 5th 2013	Notice on printing and distributing the publicity theme and slogan of 2014 China Tourism Year	National Tourism Administration	Journey of Wild China-2014 Smart Tourism Year is the publicity theme of 2014 tourism
December 2013	Technical guide on pilot project for spatiotemporal information cloud platform construction of smart cities	State Bureau of Surveying and Mapping	9 pilot cities such as Taiyuan and Guangzhou were included in the pilot project for spatiotemporal information cloud platform construction
January 9th 2014	Notice on accelerating the project of benefiting people by information	NDRC, SCOPSR, MIIT, Ministry of Finance, Ministry of Education, Ministry of Public Security, Ministry of Civil Affairs, Ministry of Human Resources and Social Security, National Health and Family Planning Commission, Audit Office, Food and Drug Administration and SAC	Gave high priority to the promotion of information consumption in nine areas, namely social security, healthcare, education, pension plans, employment, public security as well as food and drug security
January 9th 2014	2013 evaluation report on the development level of application of information technology in China	MIIT	Information level ranking of cities

(continued)

Table 4 (continued)

Date	Conferences/Files	Related organizations	Comments
January 14th 2014	Strategic cooperation agreement on the twelfth five-year plan for smart city construction	China Development Bank, MOHURD	China Development Bank will provide less than 80 billion yuan investment and financing to support the construction of Chinese smart cities in three years after the proposition of the Twelfth Five-year Plan
January 26th 2014	Blue paper of chinese international smart city development	MIIT, Xinhuanet	
February 2014	2014 ICT depth report	MIIT	100% cities above deputy provincial level were promoting the development of smart cities
March 16th 2014	National new-type urbanization plan (2014–2020)	CCCPC, State Council	Smart cities introduced to the plan formally
March 19th 2014	New-type urbanization standard system construction guideline	SAC, MOHURD	
April 22nd 2014	Announcement on the acceleration of trial project construction of intelligent urban public transport system	Ministry of Transport	Confirmed 26 cities to conduct the trial project of intelligent public transport, including Taiyuan and Shijiazhuang
June 2nd 2014	White paper of China smart grid and smart city development research report	NDRC	
June 12th 2014	Notice on the agreement on the construction of 80 national pilot cities to benefit people by information including Shenzhen	NDRC, SCOPSR, MIIT, Ministry of Finance, Ministry of Education, Ministry of Public Security, Ministry of Civil Affairs, Ministry of Human Resources and Social Security, National Health and Family Planning	Implementation of policies issued by NDRC that support smart city construction

(continued)

Table 4 (continued)

Date	Conferences/Files	Related organizations	Comments
		Commission, Audit Office, Food and Drug Administration and SAC	
August 22nd 2014	Notice of the application for national pilot smart cities in 2014	MOHURD, MOST	The third wave of application for trial started
August 27th 2014	Instructive opinions on robust development of smart cities	NDRC, MIIT, MOST, Ministry of Public Security, Ministry of Finance, Ministry of Land and Resources, MIIT, Ministry of Transport	

2.3   Categories of Intelligent City Construction Modes in China

Our intelligent city construction is mainly divided into four types, namely being oriented by intelligent infrastructure construction, by intelligent information industry construction, by service management construction for people's livelihood and by city construction featured by innovation and creativity.

1. Oriented by intelligent infrastructure construction

Cities with orientation of intelligent infrastructure construction realize detailed real-time perception of cities and increase ratio of extraction and flow rate of information by advancing network construction, raising network coverage rate, expanding network bandwidth, integrating three networks and deploying network collection facilities.

Typical cities with orientation of intelligent infrastructure construction are Chengdu, Shanghai, and Nanchang (see Table 5).

In addition, other cities with orientation of intelligent infrastructure construction include Yangzhou, Wenzhou, Fuzhou, Xiamen, Yantai, Jiangmen and Yufu, etc.

2. Oriented by intelligent information industry construction

Cities with the orientation of intelligent information industry construction start from developing intelligence relative industries, such as the internet industry to construct industrial parks, support key enterprises, issue support policies and cultivate scientific manpower as well as to use demonstrative projects to drive industrial development and then permeate into application in other areas of the city construction, such as politics and economy. Typical cities with the orientation of intelligent information industry construction are Hangzhou, Wuxi, Ningbo, etc. (see Table 6).

Table 5   Cases of cites with orientation of intelligent infrastructure construction

无线成都 智能天府	Chengdu put forward the construction concept of "wireless Chengdu" in 2009 to focus on the development of communication basis in the industry of internet of things
上海云基地 SHANGHAI CLOUD VALLEY	In the action plan of promoting the development of cloud computing in Shanghai, Shanghai emphasized that they would provide excellent basic conditions for the cloud computing needed in intelligent city construction, localize cloud computing solutions and support the intelligent city construction in Shanghai on the basis of smart technology
智慧南昌	Nanchang put "Digital Nanchang" as its emphasis in its intelligent city construction. "Digital Nanchang" will be the platform for the comprehensive command and dispatch of major infrastructure projects such as the urban intelligent traffic system, the government emergency system, "digital city transport" and "digital city management" as well as for the promotion of the operation monitoring and the level of public information service in cities to realize the strategic target of constructing intelligent cities with regional competitiveness in central China

Table 6 Cases of cities with the orientation of intelligent information industry construction

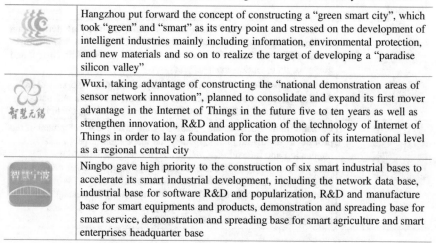

	Hangzhou put forward the concept of constructing a "green smart city", which took "green" and "smart" as its entry point and stressed on the development of intelligent industries mainly including information, environmental protection, and new materials and so on to realize the target of developing a "paradise silicon valley"
	Wuxi, taking advantage of constructing the "national demonstration areas of sensor network innovation", planned to consolidate and expand its first mover advantage in the Internet of Things in the future five to ten years as well as strengthen innovation, R&D and application of the technology of Internet of Things in order to lay a foundation for the promotion of its international level as a regional central city
	Ningbo gave high priority to the construction of six smart industrial bases to accelerate its smart industrial development, including the network data base, industrial base for software R&D and popularization, R&D and manufacture base for smart equipments and products, demonstration and spreading base for smart service, demonstration and spreading base for smart agriculture and smart enterprises headquarter base

In addition, Kunshan pointed out to strive to develop smart industries including the Internet of Things, electronic information, smart equipment, etc. Wuhan suggested perfecting the development environment of software and information service as well as accelerating the development of smart industries including information service, service outsourcing, Internet of Things and cloud computing. Cities such as Tianjin, Guangzhou and Xi'an were also constructing smart cities with the orientation of intelligent information industry construction.

3. Oriented by service management construction for people's livelihood

Cities with orientation of service management construction for people's livelihood took the development of intelligent construction closely related to city life as their entry point to carry out a large number of model application projects in the areas of public safety, urban transportation, education and healthcare, ecological environment, logistic supply chain as well as city management and to deepen smart city construction gradually from the angle closest with citizens. Typical cities with orientation of service management construction for people's livelihood are Kungshan, Chongqing, Foshan (see Table 7) and others include Beijing and Wuhan.

4. Oriented by city construction featured by innovation and creativity

The constructors of cities with orientation of city construction featured by innovation and creativity thought that creativity and innovation would be important means of developing intelligent cities, so we should strengthen urban competitiveness and realize sustainable and healthy development through the promotion of innovation in cities. Typical cities with orientation of city construction featured by innovation and creativity are Nanjing and Shenyang (see Table 8), others include Shenzhen.

Table 7 Cases of cities with orientation of service management construction for people's livelihood

	Kunshan focused on the development of intelligent transportation system, intelligent logistics, intelligent healthcare and learning e-government to promote its operational efficiency and provide more accurate control and guide for its operation management
	Chongqing pointed out to create "Healthy Chongqing" and strived to develop ecological environment, health service, healthcare and social security to improve citizens' health conditions and living quality
	Foshan Pointed out to create "Smart Foshan" with the development of ten projects as its key emphasis, namely the integration project of IT application and industrialization, strategic emerging industrial development project, rural informatization project, u-Foshan construction project, government information resource s sharing project, project for the convenience of people by IT application, urban digital management project, digital cultural industry project, e-commerce project and international cooperation expansion project

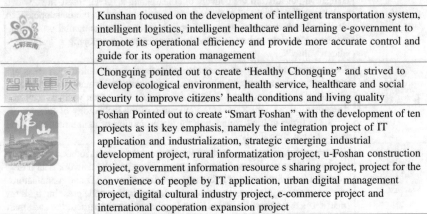

Table 8 Cases of cities with orientation of city construction featured by innovation and creativity

	Nanjing regarded "Smart Nanjing" as a vital opportunity for its urban transition. In 2012
	In the February, Nanjing issued Nanjing's Urban Development Planning for the Twelfth Five-year Plan to promote the construction of some smart industrial base projects in China like the "Wireless Valley", radio frequency valley and smart power grids base, promote the development of smart industries such as intelligent industry, intelligent agriculture, etc., motivate scientific and technological innovation with intelligent means, provide new motive force of development, stimulate industry transformation and upgrading and emphasize innovative construction so as to enhance its overall competitiveness fundamentally
	Hunnan district in Shenyang city set up the development target of the construction of a smart city with "benefiting people, promoting business and good politics" as its key words and promoted its comprehensive development capacity with strong motive force of development. At present, Hunnan is developing urban innovation through large systematic integration, interaction of physical space and cyberspace and public multi-participation and interaction to make the urban management more elaborative, the urban environment more harmonious, urban development more advanced and urban life more pleasant

On the basis of the above-mentioned construction types, devisers and constructors of smart cities should realize that the intelligent means is only an effective tool for city construction, while the ultimate goal is still to realize sustainable and elaborative urban development.

Sustainable development will be an important theoretical foundation for the development of future cities, while intelligent city construction will be the best practice of the theory.

3   Technological Support for Intelligent City Construction in China

3.1   Technological Requirement of Intelligent City System

The target of technological requirement of intelligent city system is to conduct comprehensive real-time data acquisition in cities to realize supervisory control and elaborative management in cities. Therefore, in intelligent cities, technology application is mainly reflected in the following three steps.

1. Data acquisition

Data acquisition is the first step of intelligent technology application. This step is to realize overall comprehension of people and things in cities through various sensors and data platforms to provide basic information for intelligent city construction; this step also makes up for the one-sidedness and deviation that people as individuals may make when feeling cities and lifts the level that people feel the cities. Therefore, the construction quality of data acquisition will directly influence the construction effect of intelligent cities.

The perception objects in this step involve all parts of city life, namely urban resources and environment like soil, water, vegetation, air and airspace; urban construction entity like residence, office buildings, parks and public buildings; urban infrastructure like roads, bridges, pathways, network of rivers and power grid; urban public service like education, healthcare, police affairs and real estate; and urban activities like tourism, logistics, commuting and shopping. We should take corresponding sensing technologies and means according to the types of the objects perceived.

2. Data processing

Data processing is an indispensable step between data acquisition and data application. Every day, a large number of data are produced in cities, but unprocessed data are just miscellaneous, noisy binary codes randomly arranged without any guidance functions on city construction. After processing, we can get hidden and ordered knowledge and information behind the data to make good preparations for next application.

Data analysis in this step often includes the following types.

(1) Correspondence analysis, to find relations between information from angles of space and time, etc. to find out the connection network hidden. And to add some properties that may affect the data correspondence to make the analysis result better meet demands.

(2) Similarity analysis. To find the similarity between information and divide information into several types in accordance with the similarity to find rules from a type or several different types.

(3) Predictive analysis. To establish models on the basis of rules discovered from historical data and simulate future development after on the precondition of considering future changes of influential factors. Particularly, the construction experience of large developed cities will provide a good reference for the future development of small and medium-sized cities.

(4) Error analysis. Errors will exist in the process of any date, which is unavoidable, so what we need to do is not to ignore them but to give them full consideration. We need to find the influence of errors on data processing and correct the result will greatly improve the reliability of data.

3. Data application

Data application is the last step in the application of smart technologies and is also the core part, which will have direct influence on city construction and city life.

The objects of data application can be divided into three types: governments, enterprises and individuals. Government is the serviced object on the one hand, but also a higher organizer, manager and participator in the construction of intelligent cities on the other hand, so intelligent cities will form a circulating network on the level of government. It leads the construction of intelligent cities. Meanwhile, the data feedback from it can be further used in the next round of city construction to optimize the government efficiency and elevate the development level and comprehensive strength of cities. Enterprises can propel city development, so the application of enterprise data will optimize business models and results of enterprises, feed city development, promote the development of intelligent related enterprises and elevate the intelligent level of cities. On the personal level, individuals usually benefit from receiving data passively. In fact, this level can reflect the degree of intelligence cities to the largest extent. To improve citizens' living standard through intelligent means is one of the important targets of intelligent city construction.

3.2   Critical Technologies in Intelligent City System

The range of technologies for intelligent city construction is so wide that it is hard to name each one, so we will only introduce some core technologies in intelligent city construction below.

1. Internet

Internet is the basic network of open telecommunication to provide high quality telecommunication services with broadband IP as the core technology. In recent years, with the advancement of telegraphy, Internet in China has developed rapidly.

By the end of June 2014, netizens in China totaled 632 million and the coverage rate of Internet was 45.9%, among which 527 million were using mobile phones to access to the Internet (China Internet Network Information Center 2014). In fact, the internet construction is still insufficient in China. Our average internet speed is only 3.8 Mbs, ranked 75 in the world and the access speed lags far behind the US, Japan and other Internet developed countries (Belson 2014).

The Twelfth Five-year Plan of Broadband Network Infrastructure suggests that fiber broadband network promotion project will be one of the five major projects in our broadband network infrastructure construction during the Twelfth Five-year Plan period. During the Twelfth Five-year Plan period, newly built areas in cities mainly applied the method of Fiber to the Home (FTTH) and were covered by wired broadband network with FTTH rate reached over 60% in newly built houses in cities. Optical access was promoted for replacing copper access in constructed areas in cities to accelerate network transformation.

In recent years, the development of mobile internet is gaining momentum, which has combined mobile communication with internet and is more convenient and portable compared with average broadband network. The development of mobile internet changed from 1G network to 4G network. Compared with developed countries, mobile network construction in China is still in its initial stage and problems such as low network coverage, low quality of network service, slow network speed and high user charge, etc. are affecting the advancement of intelligent cities, but with policy support and industrial development, our mobile network will be sure to achieve higher targets.

2. Internet of Things

As for "what is Internet of Things on earth?", different countries have different opinions.

A common definition is: it is a network that can connect any things with internet and conduct information exchange and communication according to common agreements through sensors such as radio frequency identification devices (RFID), infrared sensors, global positioning system and laser scanner, in order to intelligent recognition, localization, tracing, supervision and management (Li 2009).

However, the definition of Internet of Things in the EU is: Internet of Things is a part of future internet, which can be defined as a dynamic global facility network with the ability of self-configuration on the basis of standardized and interactive communication agreement. In the Internet of Things, physical and virtual "objects" have characteristics such as identities, physical properties and personification, which can be connected by an integrated information network.

With the effects of Internet of Things gradually being understood by people, all regions in China begin to issue plans of Internet of Things and compete to develop industries in this regard. However, Liu Yunjie, an academician of China Engineering Academy showed his worries on the summit forum for the future development of electronic technologies and cultivation of relative talents on December 9th, 2011, "In my point of view, the focus of the application of Internet of Things should be to overcome challenges met in national welfare and people's livelihood as well as social development. It will be meaningless if we purely pursue profits but ignore the above demands or just follow the trend blindly".

3. Cloud computing

Cloud computing is the development and commercial application of concepts such as distributed processing, parallel computing and grid computing with virtualization of IT software and hardware resources like computing, storing, server and utility software as its technological nature (Ma 2012). The State Council, MIIT, NDRC and other relative government sectors all stressed the importance of developing cloud service innovation and application demonstration in Decisions of the State Council on Speeding up the Cultivation and Development of Strategic Emerging Industry and Notice on Conducting Pilot and Demonstration Work of Cloud Computing Service Innovation and Development. At present, there have been over twenty cities issuing estate plans and project plans related to cloud computing and promoting the construction of such cloud computing infrastructure as Internet Data Center (IDC) and Disaster Recovery Center through the paths of government- enterprise cooperation or industry-university-research cooperation; at the same time, enterprises such as Alibaba and Baidu are constantly enriching the application of cloud computing in their practice and optimizing users' usage experience.

4. Space information network technique

Space information network is composed of satellites and constellations of different orbits, types and properties that connect satellites or satellite and ground on the one hand and corresponding ground facilities on the other hand, and it is also a collection of command, control, communication and other application systems supported by space information network. Space information network is the space segment in the "space-to-ground" integrated network and is composed of communication infrastructure, network infrastructure and network application facilities, which covers the global communication and realizes seamless connectivity of heterogeneous networks to support different applications (Pan 2014).

Satellite Positioning, Remote Sensing and Geographic Information System (GIS) that we currently use all belong to the range of researches on space information network technology. They collect, store, manage, analyze and apply space information data distributed geographically through combination with computer technology and communication technology.

5. Data consolidation technology

The concept of data consolidation which includes system integration, application integration, host integration, storage integration, database integration and data centralization, is relatively confused. These concepts are all connotation and denotation of computer system integration from different levels and angles. In fact, consolidation is an exotic vocabulary, meaning to merge, solidify and intensify. Its original meaning is to do comprehensive construction on the original base not to make a new start or full update.

The process of intelligent city construction in China involves collection and processing of a lot of data. However, in the process of information construction in

China, due to differences in such aspects as source, time, equipment and platform, there is a large amount of redundancy and inconsistency between batches of data, which brings great difficulties to integrated application of data. It also leads to the situation where existing information cannot be used. Besides, isolations of information are multi-dimensional, which means the elimination of these isolations will be complex and time-consuming.

Therefore, data consolidation technology is indispensable and vital in intelligent city construction. The technology is especially professional and closely related with actual business, which means designing programming and integrity process without prospective will certainly derive larger islands while removing some existing ones. It needs not only the technology itself to be mature gradually, but also standardization in the national level as soon as possible.

4   Trends of Intelligent City Construction in China

4.1   China Intelligent City Construction Entering a New Phase

According to the statistics from the 2014 ICT In-depth Report of MIIT, 100% cities above the deputy provincial level, 89% cities (241 cities) above the prefecture level and 47% cities (51 cities) above the county level are promoting the smart city construction. National pilot smart cities published by MIIT totaled 193. On the regional level, cities above the prefecture level listed in the 2013 government work repot or the local Twelfth Five-year plan to be the key cities in smart city construction reaches 52 in number.

In March, 2014, National New-type Urbanization Plan (2014–2020) issued by the State Council put forward clear demands on promoting smart city construction, comprehensively using material resources, information resources and intellectual resources of urban development, stimulating creative application of new generation information technology like Internet of Things, cloud computing and big data as well as realizing deep integration with the urban economic and social development, thus drawing a route map for the development of smart cities in China.

2014 Smart City Forum was hosted in Beijing on May 15th, 2014 with the theme "Open, Development and Service—Focusing on the New Stage of Smart City Construction". At the forum, based on the current situation of our intelligent city development, experts supposed that 2014 would be the "First Year of Smart City Landing". Intelligent city construction in China has turned from the pilot stage featured by doctrinarism and imagination into the practical stage featured by down-to-earth planning. Meanwhile, the mature development of technology, opening and sharing of information and orderly revolution of management, etc. will all speed up the intelligent city construction.

According to the research of IDC, an international data corporation, the capacity of the IT market in 2012 reached 9.3 billion dollars, among which solutions such as digital urban management, smart healthcare and smart transport account for a large proportion. Stimulated by the investment of 80 billion RMB on smart city market by the China Development Bank and MOHURD, the capacity of the market enhanced substantially in 2013 compared with that in 2012, increasing 18.5%. Markets with solutions like public security, digital urban management, smart healthcare and smart transport increased especially rapidly, among which markets like IT service, software, intelligent terminal and business PC remain relative a high growth rate.

4.2   *China Intelligent City Construction at Primary Stage*

Intelligent city construction in China is growing vigorously, and the construction of many cities involves many concepts such as Internet of Things, ubiquitous network and cloud platform, which have been interpreted by experts from various angles.

It is undeniable that most people still lack deep understanding about how these terms should be applied in specific construction, so eventually, all constructions tend to be similar and lack breakthrough. Therefore, we still need to analyze in combination with practical problems and thoroughly understand the importance of these terms on urban transition and development.

The intelligent city evaluation system needs to be perfected and meets the governmental evaluation system. Traditional evaluation methods of city construction based on GDP and industrial development are obviously not suitable for smart cities, for they cannot make response and judgement on the construction effect that smart city should have.

At present, there are many evaluation systems of smart system and many regions in China are exploring in this field as well, but the current existing system is still in need of perfection and should be risen to the national standard to meet the fast pace of city construction so as to be an influential factor in deciding urban development policies.

Information should be more open to the public to get through all information channels. Government often masters the most detailed and comprehensive data in various types of the whole society, from city operation to personal income, but what matters most is the opening of information if we truly want to construct intelligent cities and develop information economy. On the one hand, the government needs to grant people the "right to know" to realize information transparency; on the other hand, the most important thing is to guarantee relative organizations' and individuals' right to use information, which can then make sure that these data can be made the best use in the process of intelligent city construction.

Departments in the government definitely should form an effective information interchange system to avoid repetitive work and information contradiction and ensure effective formulation and implementation of various policies. But now we still open information in a passive way and governments lack the concept of active information disclosure. Besides, because of departmental benefits and confidential demands, different departments have their own independent databases. Governments can consider about finding an entry point of intelligent city construction in terms of enterprise cooperation to give priority to the development of service-oriented industries. For example, Shanghai has established demographic database, space geographical database and enterprise legal entity database, which have laid sound foundation for sharing and excavating "concentrate" crossing departments and industries.

4.3 Future Steps for China Intelligent City Construction

Chinese Academy of Engineering has listed the following 12 key points in intelligent city construction: urban economy, science and technology, culture, education and management, urban spatial organization pattern and intelligent transport and logistics, intelligent power grid and energy grid, intelligent manufacturing and design, knowledge center and information processing, intelligent information network, intelligent building and furnishing, intelligent healthcare, urban safety, urban environment, intelligent business and finance, and spatio-temporal information infrastructure in intelligent cities. The 12 points can be categorized into five parts: intelligent urban construction, intelligent urban information infrastructure, intelligent urban industrial development, intelligent urban management service and intelligent urban human resources. All the parts make up the construction of intelligent cities. Given that our intelligent city construction is still at primary stage with many problems and challenges to be overcome, intelligent city construction in China can be promoted in three steps gradually.

1. Unit construction

The first step is to start from intelligent industries and single intelligent unit construction, such as intelligent transportation, intelligent healthcare and intelligent community. These units are important parts of intelligent cities, but starting from single unit will limit the range of researches, facilitate the depth of researches and ensure the results of the single unit construction. This is the common stage which our practice is at now.

2. Regional construction

The second step is to start with large intelligent residential areas or urban intelligent districts. On the one hand, we should organize the above units comprehensively and apply them in a certain urban area to verify the effects of the intelligent system when solving real urban problems; on the other hand, the cooperation between different industries and departments shall be tested. Comprehensive regional construction will lay sound foundation for the popularization of intelligent city construction in the whole city and the whole nation at large.

3. Overall construction

The third step is to put intelligent city construction into practice extensively. After accumulation and verification in the first two phases, people will have a deeper understanding of various aspects of intelligent cities and will have recognition of and counterplan for obstacles in local intelligent construction. Besides, the development and application of technologies tend to be mature and the linkage among individuals, enterprises and governments will pass through the breaking-in period. At this stage, promoting large-scale intelligent city construction will be reasonable and it will elevate the construction efficiency and operation efficiency of intelligent cities.

References

Belson D (2014) State of the internet: Q3 2014. https://www.akamai.com/us/en/multimedia/documents/content/akamai-state-of-the-internet-report-q3-2014.pdf. Accessed 10 July 2015

Chen D (2014) Policy documents on intelligent city compilation. http://www.ccud.org.cn/2014-11-03/114354172.html. Accessed 11 Dec 2015

China Internet Network Information Center (2014) 34th China internet network development state statistic report. http://www.cnnic.net.cn/hlwfzyj/hlwxzbg/hlwtjbg/201407/t20140721_47437.htm. Accessed 20 Oct 2014

Li F (2009) 1 principal line, 6 changes, 7 key points and 4 measures—interpret China radio frequency identification (RFID) technology development report blue book. China Autom Ident Technol 6:41–42

Ma YC (2012) Research and implementation on resource management facing cloud market. Jiangsu University of Science and Technology, Zhenjiang

Pan CS (2014) Several critical techniques on spatial information network. Commun CCF 9(4):46–51

Sun LX (2014) The overall trend and core concept thinking of intelligent city construction. China Inf 2:84–87

Yang BZ (2012) Six construction characteristics and four kinds of operating modes of intelligent city construction in China. http://www.im2m.com.cn/107/10125065454.shtml. Accessed 29 Nov 2015

Zhou P (2013) Intelligent city: the leading trend of city development. Inf Const 8:38–40

Zhou WP (2014) The present situation of intelligent City in the whole world [EB/OL]. http://www.istis.sh.cn/list/list.asp?id=8033. Accessed 1 Dec 2015

Chapter 3
Development Stages of Intelligent Cities

Abstract Throughout the history of urban development, along with the continuous development of science and technology, urban development has undergone three social civilization stages: agricultural civilization, industrial civilization and information civilization. Intelligence is considered as a product generated in the constant in-depth development of information civilization. Although relevant research areas of intelligent city have recently attracted lots of attention, researches on the development stages over time are relatively lacked. The significance of this chapter is to sort out the development stage of smart city at home and abroad, and explain the law of the development and evolution of smart city. In this book, the development of intelligent city practices are divided into three stages: stage I, empirical preparation and the formation of conception, stage II, dissemination of intelligent city conception and experimental practices and stage III, in-depth exploration. For each stage, main features and connotation, landmark events and prototypical practice cases are identified.

Keywords Intelligent city · Development stage

Throughout the history of urban development, along with the continuous development of science and technology, urban development has undergone three social civilization stages: agricultural civilization, industrial civilization and information civilization.

In Chinese, the word that is equivalent to "city" is made up of two characters, cheng and shi. "Cheng" refers to walls built to defend a city, while "shi" refers to a fixed site for trade. Cities in the era of agricultural civilization embody the two functions mentioned above, and often play the role of the military, agricultural, economic and political center. In 1840, the Industrial Revolution led the cities into industrial civilization stage and the economy had undergone epoch-making changes. With a fast gathering of production factors and population in cities, and the mutual promotion of industrialization and urbanization, cities expanded and developed rapidly and the productivity experienced an unprecedented increase on the basis of the agglomeration development effect of cities. As the information

© Springer Nature Singapore Pte Ltd. and Zhejiang University Press 2018   45
Z. Wu, *Intelligent City Evaluation System*, Strategic Research
on Construction and Promotion of China's Intelligent Cities,
https://doi.org/10.1007/978-981-10-5939-1_3

technology represented by the Internet continuously develops to maturation, the urban development gradually enters information society stage. Human beings have gradually deepened their perception on the objective law of urban development with the assistance of information technologies. The operation of cities becomes more readable, visible and controllable, which indicates that cities are stepping into the efficiency optimization stage of urban development.

Intelligence, however, is a product generated in the constant in-depth development of information civilization when the information civilization comes to a certain stage. The core idea is to consider a city as an intelligent life that has its own neural network and independent decision-making feedback system, and also has the ability to learn and to evolve (Shi and Li 2006).

Although relevant research areas of intelligent city have recently attracted lots of attention, such concept has only been brought up for several years. Both domestic and foreign scholars are more inclined to do research in terms of categories and patterns of practice content of intelligent city, while researches on the development stages over time are relatively lacked. Here we take the research of Wang (2012) as an example. In his study, he divided the development of intelligent city into three stages based on the construction process of intelligent city, and these three stages are pre-intelligent city construction stage, primary stage of intelligent city construction and advanced stage of intelligent city construction. The focuses of the three stages are respectively on technological development, network and service construction, and an advanced status of integrated society, environment and management. However, it is a regret that no further demonstration combined with domestic and foreign practices has been provided in his study. Peng (2012) has conducted researches on categories, in which he analyzed practice cases at home and abroad such as Stockholm, Shanghai, Ningbo, etc. and did classification based on subjects of construction combined with the development background and operation status of each practice. As a result, he found that disparities in categories of intelligent city were mainly determined by their development backgrounds, without any obvious staging characteristics.

Based on the emphasis of China's intelligent city construction, Xu (2012) classified the construction patterns of intelligent city into three categories, which are the pattern oriented by information infrastructure construction, the pattern driven by development of Internet of Things industry and the pattern supported by social services and management application.

Although intelligent city practices at home and abroad have breakthroughs in several areas, it still needs continuous exploration and experimentation before being developed to maturity. Therefore, we have divided intelligent city practices into three development stages, which are stage I, empirical preparation and the formation of conception, stage II, dissemination of intelligent city conception and experimental practices and stage III, in-depth exploration. Further, combined with practice cases at home and abroad, we explain the connotation of each stage by using the emphasis and features in the process of constructing intelligent city to analyze main features and connotation, landmark events and prototypical practice cases in each stage.

1   Stage I: Empirical Preparation and the Formation of Conception

1.1   Features and Connotation of Development Stage I

Stage I: Empirical Preparation and the Formation of Conception are from 1990s to the year 2009, which includes the early foundation for the conception of intelligent city and the presentation of intelligent city. The conception of smart city was presented for the first time in 2008, but a wide range of previous practices for related concepts such as city informatization and digital city have laid a solid foundation for the formation of intelligent city conception. Thus, the conception of intelligent city has attracted significant attention and discussion from all over the world since it was formed. The practice of intelligent city was ready and waiting.

The conception "intelligent city" is a deepening of "city informatization" and "digital city". In 1980s, developed countries had gradually transformed from industrial society into information society. After several years of preparation, China launched a few major informatization projects in 1993. This is the beginning of city informatization, with a major aim of information infrastructure construction, such as laying optical fiber, transnational connection of submarine cable and expansion of bandwidth (Yang 2003). In 1993, Albert Arnold Gore Jr., the then US vice president, firstly used the term "digital city" in his report "The Digital Earth: Understanding our planet in the 21st Century" (Mi 2001). Later, China conducted construction projects of digital city on a nationwide scale, the core of which were to deal with urban issues through digital measures, and to completely integrate and utilize information resources on the basis of information infrastructures, enabling the sharing, circulation and interoperability of a city's information.

On the basis of the previous theories and practices and by using a new generation of information and communication technologies, the creation of a self-sensing, analyzing and decision feedback system for urban information networks is emphasized in the conception of intelligent city.

The essence of such concepts is actually the breakthrough innovation and development of information technology. From 1980s to 90s, technical innovation in personal computers, data storage and process as well as network communication equipment's had gradually propelled information trend of cities. In 1997, the popularity and promotion of information technology, represented by the Internet and application software, enabled the information transmission and communication of human beings to break through the limit of time and space, leading the practice of digital city concept. On the other hand, the third information technology revolution represented by the Internet of Things and cloud computing symbolizes the advent of an era for intelligent interconnection of city information. In 2006, some scholars have observantly made predictions based on the maturity of the Internet of Things technology development that digital city will evolve to a higher stage of intelligent city (as shown in Fig. 1).

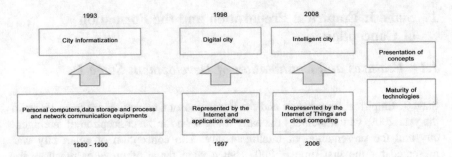

Fig. 1 Interaction between maturity of technologies and presentation of concepts

1.2 Landmark Events

In 2008, Samuel Palmisano, CEO of IBM, put forward the conception of "Smart Planet" for the first time in his report "A Smarter Planet: The Next Leadership Agenda". This is the landmark event in this stage. The core of the conception is to become more instrumented, interconnected and intelligent. At the end of 2009, IBM corporation brought up the conception "smart city" and its solutions on the basis of "Smarter Planet". Since then, thanks to the active avocation and promotion of IBM, the conception of "smart city" had attracted great attention and high recognition from the society. It has gradually been accepted and studied in major developed countries and regions in the world, which starts the construction boom of smart city.

1.3 Prototypical Practice Cases

In these stages, a new generation of information and communication technology has been applied consciously in practice cases. With the application of innovative information technology, it is more efficient to achieve all urban development aims, such as industrial economy development, social management, etc. But the practical theory of smart city has not been fully reflected. Limitations in this stage are that application scope of new technologies is restricted and has not been promoted to city, development visions are not considered from the perspective of smart city during practice and the aims are always optimization of local system efficiency.

1. Domestic practice

In 2009, Premier Wen Jiabao pointed out that we shall establish an information sensing center or the "Sensing China" center to compete in the fierce international competition during his visit to Wuxi. On November 3, 2009, Premier Wen Jiabao made a speech titled with Let Science and Technology Lead Sustainable Development in China, in which he pointed out that we shall make every effort to master the key technology of sensing network and Internet of Things, so that the information and network industry would become an "engine" to drive industrial upgrading and advance to information society. Meanwhile, Internet of Things became one of the six strategic emerging industries. Soon China started to construct intelligent city and a trend to develop Internet of Things industry (Qiu 2013).

2. International practice

In 2006, EU established an organization called Living Lab. It has been more convenient to transmit information and share knowledge with the application of a new generation of information and communication technologies. Living Lab tried to challenge conventional science and technology innovation activities led by researchers and carried out in laboratories with innovative activities based on knowledge to inspire the wisdom and creativity of all communities and encourage all citizens to participate in science and technology innovation.

2   Stage II: Dissemination of Intelligent City Conception and Experimental Practices

2.1   Features and Connotation of Development Stage II

Stage II: Dissemination of Intelligent City Conception and Experimental Practices is from 2009 to 2011. During this period, the strategic significance of intelligent city development had been widely accepted and quickly recognized after the empirical preparation and the formation of conception in the previous stage. In January 2009, Obama, after his inauguration, responded positively to the conception "Smarter Planet" put forward by IBM CEO Samuel Palmisano and made it a national strategy (Zheng 2011).

In 2009, IBM released the report "Wisdom Earth Win in China" and white paper "Smarter City in China" to bring up that the ultimate goal and strategic direction of city informatization should be to establish intelligent city. During World Expo held in Shanghai in 2010, IBM gave a speech with the topic "Constructing a Smarter Earth from Cities" in intelligent city global summit held in Shanghai, to disseminate the conception of smart city towards all communities, which has aroused great attention and widespread concern.

Since then, the planning and deployment of intelligent city practice started formally both in China and in other countries. Several cities took the lead in practice and exploration. As the background of intelligent city concept is the maturity of key technologies such as Internet of Things, the practical exploration in this stage is strongly technically oriented, and its connotation is more focused on the intelligence level of information technology.

The essence of all practice cases in this stage is trying to apply the new generation of information and communication technologies that are increasingly mature in all fields of the city under the guidance of intelligent city theories, mainly aimed at improving the effectiveness of all systems in a city. Comparing with practices in the first stage, the most significant difference in this stage is the preliminary application of intelligent city conception and practice theory. For example, the goal to be more instrumented, interconnected and intelligent has been reflected in part of the practice cases. Thanks to the widespread intelligent city concept, the number and range of practice cases in this stage have experienced a rapid development.

2.2 Landmark Events

The practice attempt of intelligent city in 2010 Shanghai World Expo was the landmark event of this period. Information and intelligent technologies have been fully applied in the planning, construction, operation during exhibition and post exhibition utilization and development of Shanghai World Expo parks, with proactive planning of information and communication infrastructures to offer a high level of communication and information service. With the construction of nearly 40 information application systems and the integration with operation of the World Expo, a visible, controllable and sustainable development has been achieved and the operation and management of the parks have been improved efficiently. The examples are as follows: an intelligent system for traffic information management offering real time monitoring and feedback for passenger flow and traffic information in and outside of the parks; an energy and environment system monitoring indexes such as electric power, water supply, humidity and temperature, noise, etc. (Cheng 2011). This attempt at Shanghai World Expo symbolized that the concept of intelligent city had been spread and concerned widely. The practice stage is coming.

2.3 Prototypical Practice Cases

In this stage, the new generation of information and communication technologies are gradually applied and attempted in practice cases under the guidance of intelligent city theories. Specifically, in each practice case, the key features of intelligent city theories have been highly valued, such as being instrumented, interconnected and intelligent feedback. However, in this stage, there are some differences between the practices inside and outside China. Domestic practices pay more attention to macro design of strategies, with a large quantity and wide range, while foreign practice are more prudent as a certain field are often chosen to be a sally port.

1. Domestic practice

In 2011, An Action Plan for Shanghai to Promote the Construction of Intelligent City (2011–2013) was proposed in Shanghai, which was the beginning of intelligent city construction. Specific aims and tasks included the basic construction of broadband city and wireless city, preliminary presentation of the efficacy of information sensing and intelligent applications, new information technology industry becoming a powerful support for the development of intelligent city, information security generally being trustful, reliable, and controllable.

The construction of intelligent city in Taiwan had a sufficient practice preparation, including Development Plan of Information and Communication implemented in 1988, the "e-Taiwan" plan developed during 2002–2007, "M-Taiwan" plan developed during 2005–2009, which formed a good foundation of information technology with a good popularity rate. In 2009, Taiwan launched "Intelligent Taiwan" plan. Characterized with being interconnected and instrumented, a smooth internetwork has been established by upgrading high speed broadband network. Combined with the setup of sense networks, a perfect broadband infrastructure and application environment have been built (Yan 2013).

2. International practice

In 2009, Dubuque of Iowa and IBM announced to build the first smart city in the US. With the application of the Internet of Things technology, public resources in the city such as water, electricity and gas were monitored, analyzed, integrated and released automatically through the installation of household digital control water and electricity meter and the establishment of a comprehensive monitoring platform, which had a relatively complete practice of the smart city theories.

3   Stage III: In-Depth Exploration

3.1   Features and Connotation of Development Stage III

Stage III: In-depth Exploration starts from 2011 and continues up to now. In this stage, after the dissemination of intelligent city conception and the practices accumulated in the previous stages, the exploration of intelligent city practices gradually deepens: on the one hand, the quantity, range and scale of the practice increase rapidly; on the other hand, the connotation of smart city is constantly enriched. Scope of practice and scale of development are the focus in domestic development. In the 12th five-year plan released in 2011, there are more than 20 cities that listed smart city in the construction plan, including Beijing, Tianjin, Shanghai, etc. However, foreign practices have gradually transited from technology orientation to multi-orientation, emphasizing the optimization and advance of the urban and social development, and paying more attention to the intelligence of the overall society instead of just technology. Taking Amsterdam as an example, smart city practices in this city have a theme—sustainable development, aiming at reducing 40% of greenhouse gas emissions by various innovations and technologies.

The development of this stage is based on the initial attempt and exploration of domestic and foreign cities. Due to different development basis and demand of each city, technology-oriented smart cities have large practice limitations, thus some cities started to seek for their own smart city practice pattern. The connotation and applicability of the smart city practice have been greatly improved at this stage, such as the integration of sustainable development, innovative city and other development concept. As a result, the field of practice has also been rapidly expanded.

3.2   Landmark Events

At this stage, the landmark event is that the Ministry of Housing and Urban-Rural Development announced two batches of national smart city pilot lists in 2013. In January 2013, the Ministry of Housing and Urban-Rural Development released the first batch of national smart cities, including 90 pilot projects. The same year in August, the Ministry of Housing and Urban-Rural Development released the second batch of national smart cities, including 103 pilot projects. As of the end of 2014, there were 193 pilot projects for national smart cities, which indicated that China's smart city practice had reached a comprehensive and in-depth practical exploration stage.

3.3   Prototypical Practice Cases

At this stage, domestic practices are more focused on expanding the scope and scale of practice, while foreign practices have gradually transited from technology orientation to multi-orientation.

1. Domestic practice

In 2011, after the completion of preparation of "Outline of the Action Plan for Ningbo City to Speed Up the Construction of Intelligent City (2011–2015)" and other planning preparations, Ningbo officially launched the practice of Ningbo smart city. The practice was conducted on a large scale, with a wide range and rapid progress. Specific actions included: identifying 61 intelligent application system demonstration projects, signing a smart city cooperation project with a total investment of about 6.5 billion yuan in the first China Intelligent City Technology and Application Products Expo, investing more than 5 billion yuan to build 30 major projects of intelligent city construction and 19 major project constructions of intelligent industry.

2. International practice

In December 2013, the London City Council released the "Smart London Plan", focusing on the opportunities and challenges that London will encounter in 2020, and improving the competitiveness of London's cities and the quality of life through innovative technology. The plan put forward seven directions for the construction of smart city in London, including public life services, public data, innovative talents, innovative enterprises, and each direction has put forward the corresponding action policies and specific indicators and requirements for development goals.

4 Conclusion

"Smart city" is not a modern concept sold by packaging and propaganda. The significance of this chapter is to sort out the development stage of smart city at home and abroad, and explain the law of the development and evolution of smart city. This chapter answers the following three key questions about the theory of smart city development.

(1) The presentation of smart city concept has its background, which is consistent with the development trend of information technology and has a wide range of practical basis. Therefore, smart city concept can be widely recognized and arouse a trend for practice in the world. On the one hand, after the city informatization, practice and exploration of the digital city concept, information technology has been popular to a certain extent, laying the foundation for information technology practice; on the other hand, due to the breakthrough and innovative development of information technology, new conceptions of urban development have been put forward, and application of Internet of Things and cloud computing technology have promoted the formation of smart city conception.

(2) The essence of smart city practice cases determines the stage of its development. The application of innovative information technology is a common feature of all practice cases. The characteristics of intelligent city practice at different stages are shown in Table 1. The empirical preparation and the formation of conception stage are based on the practice of information technology. The practice under the guidance of smart city theory provide the underlying basis for dissemination of intelligent city conception and experimental practices. Seeking for practice pattern of smart city to meet their own demands is the core of in-depth exploration stage.

(3) The essence for the evolution of smart city is the change of driving forces of development throughout all development stages. The evolution process is first to promote the formation of conceptions through technological innovation, to promote the initial practice with visions and assumptions of the theory, and finally reaches the stage where city goes into in-depth exploration stage based on their own development demands. Landmark events in the development stage of intelligent city are shown in Fig. 2.

Table 1 Characteristics of intelligent city practice at different stages

Development stage		Empirical preparation and the formation of conception	Dissemination of intelligent city conception and experimental practices	In-depth exploration
Time range		1990s–2009	2009–2011	2011–Now
Development characteristics		Practice of information technology	Attaching great importance to smart city theory	The quantity, range and scale of the practice increase rapidly; the connotation of smart city is constantly enriched
Essence		Breakthrough and innovative development of information technology	Practice under the guidance of smart city theory	A practice pattern of smart city to seek to adapt to their own demands
Landmark events		IBM put forward the conception of smart city for the first time in 2008	The practice attempt of smart city in 2010 Shanghai World Expo	Ministry of housing and urban-rural development announced two batches of national smart city pilot lists in 2013
Prototypical cases	Domestic cases	Internet of things became one of the six strategic emerging industries	Smart city construction in Shanghai	Smart city practice in Ningbo
	International cases	LivingLab organization in Europe	Smart city practice in Amsterdam	Smart city practice in London
Driving forces of development		Technology	Vision	Demand

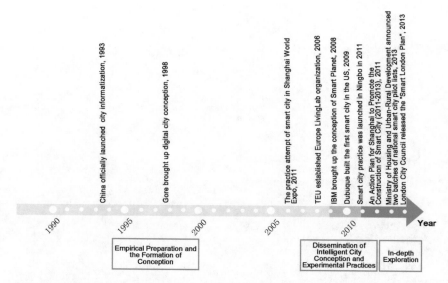

Fig. 2 Landmark events in the development stage of intelligent city

References

Cheng DZ (2011) Explore intelligent cities—the inspiration of Shanghai World Expo 2010 in China. In: City development research—collected papers of the 7th international conference on green building and building energy efficiency

Mi W (2001) The concept and historical background of digital city. City Plann Commun 17:10

Peng JD (2012) Research on the mode of intelligent city construction in the world. Jilin University, Jilin

Qiu BX (2013) China intelligent city development research report (2012–2013). China Architecture & Building Press, Beijing

Shi WY, Li Q (2006) Digital city: the primary stage of Intelligent city. Earth Sci Front 13(3):99–103

Wang SW (2012) Talking about "intelligent city". Libr Inf Serv 56(2):5–9

Xu JH (2012) Current situation and type comparison research of intelligent city in China. City Obs 4:5–18

Yan BB (2013) Intelligent city construction experience and enlightenment—taking Taoyuan, Taiwan as example. Contemp Econ 11:41–43

Yang XW (2003) City informatization research in the process of modernization in China. Central China Normal University, Wuhan

Zheng LM (2011) Strategic thinking about intelligent city construction. Mod Manag Sci 8:66–68

Figure ...

References

Chang, D... (2011) ....
....
....

Ali, M... (2012) ....
....

Di, B... (2011) ....
....

Tan, C... (2012) ....
....

Wang, G... (2013) ....

Wang, H... (2013) ....
....

Yu, M... (2011) ....
....

Yang, X... (2010) ....
....

Zhang, J... (2012) ....

Chapter 4
Evaluation Indicator System

Abstract Indicator System refers to an organic entity composed of several statistical indicators that are interconnected. The establishment of the indicator system is the prerequisite and basis for the prediction or evaluation study. The conception of city indicator system is the extension of academic conception of the indicator system. From the view of system composition (i.e., attribute of the selected indicator) and the structure of the system (the way each part is organized), this chapter summarizes the internal logic of the urban evaluation indicator system at present from four aspects: system composition, system structure, indicator calculation and system calculation. Moreover, key characteristics of city indicator systems and main traps in the structuring city indicator systems are identified in order to ensure the objectivity and realistic operability of evaluation indicator systems.

Keywords Evaluation indicator system · Urban indicator

1 City Indicators

From the view of statistics, research design and social sciences, an "indicator" is the behavior based on the representation or implicit characteristics of a research object. As an important tool for quantitative research, indicators are widely used in natural and social sciences.

An urban indicator is behavior of the representation or implied characteristics obtained by taking the city as a research object. With the global urbanization rate exceeding 50%,[1] the city has become the most important place for human life. As windows to understand the city, these indicators become the most important quantitative tools for urban research.

The most significant feature of urban indicators is a city's macroscopic characteristics and richness. Generally, urban indicators can be classified into social, economic, political, cultural, ecological and other categories according to their

---

[1]For details, see http://wdi.worldbank.org/table/3.12.

© Springer Nature Singapore Pte Ltd. and Zhejiang University Press 2018

Z. Wu, *Intelligent City Evaluation System*, Strategic Research
on Construction and Promotion of China's Intelligent Cities,
https://doi.org/10.1007/978-981-10-5939-1_4

attributes. Urban indicators can also be classified as ones directly available and ones available after being reworked in terms of technical access.

The contribution of urban indicators can be interpreted as: ① urban indicators can reflect the performance of a city in terms of the attribute of a single category; ② urban indicators can be used for horizontal comparison between cities, forming a comparison relationship of such attribute between cities; ③ urban indicators are extensible, and the indicator level is enriched with the combination of indicators. Multiple indicators can represent a phenomenon or problem in a complex way; ④ when urban indicators form a system reflecting the evaluation function, a large number of urban evaluation indicator systems will be formed. As an important indicator collection to measure of the overall operation of a city or performance of specific aspects, urban evaluation indicator system has a scientific and operable characteristic.

2   City Indicator System

2.1   Conception of City Indicator System

Indicator System refers to an organic entity composed of several statistical indicators that are interconnected. The establishment of the indicator system is the prerequisite and basis for the prediction or evaluation study. It is the process of decomposing the abstract object into a behavioral, operable structure according to its essential attributes and identity of certain features, and then giving each constituent element (i.e., indicators) in the indicator the corresponding weight. The conception of city indicator system is the extension of academic conception of the indicator system. Although it is not clearly defined, there is already a broad consensus in the application field.

The common city indicator system can be classified into four categories: ecological-environment evaluation system, social-economic evaluation system, information evaluation system and policy-management evaluation system. On this basis, there are many evaluation indicator systems which cross with these four categories, generally based on the evaluation of urban development and operation-specific projects, such as the sustainability of urban development, the intelligent level of urban construction and the degree of civilization of the city.

Evaluation systems that cross with multiple categories include: Energy and Sustainable Urban Evaluation System (Xu and Zhang 2011), Urban Land Resource Sustainability Evaluation System (Tan et al. 2003), Intelligent City Evaluation System (Fu 2003), Legal Urban Evaluation System (Weng 2011), Eco-civilized city Evaluation Indicator System (Guan et al. 2007), Environment-friendly City Evaluation Indicator System (Wen and Li 2007). Although there are differences between the construction of these evaluation systems and common evaluation system, the selected indicators can be grouped into common types based on their attributes.

In each type of city evaluation indicator system, the city indicators are constantly screened, combined and calculated in a compounded way to finally form a collection of important evaluation tools that serve different evaluation indicator systems.

From the view of system composition (i.e., attribute of the selected indicator) and the structure of the system (the way each part is organized), the internal logic of the urban evaluation indicator system at present can be summed up from four aspects: system composition, system structure, indicator calculation and system calculation.

2.2 Composition of City Indicator System: Combination of Indicators with Different Attributes

From the attribute of the index itself, it is possible to obtain the composition of different evaluation indicator systems more clearly by classifying all the indicators involved in all the systems

(1) Social and economic indicators, including demographic data, per capita income data, regional income data, fiscal data, Gini coefficient, and so on. Most of the evaluation indicator systems have chosen to obtain basic information from such indicator collection as an integral part of the evaluation objectives or as a review indicator for the calculation of other parts.
(2) Environmental and ecological indicators, including air quality data, water quality data, noise index, species diversity data, etc. Indicators in this collection are often used in the evaluation indicator system involved in sustainable development and environmental issues.
(3) Infrastructure development indicators, including traffic data, information engineering data, energy data, and so on. Infrastructure mentioned here is different from the social construction contents relating to social development in social and economic indicators. The hardware system needed for urban development is required here, as well as the degree of informatization of such hardware facilities.
(4) Governance and management indicators, including fiscal transparency, corruption data, policy implementation assessment data, etc. This data collection is mainly a series of evaluation data obtained by the city government in the process of urban development and construction to evaluate the effect of governance and management. As for specific evaluation indicator system, relevant governance performance indicators of the government will be selected and included in the evaluation system.

To understand the urban evaluation index system structure, a comprehensive analysis needs to be carried out in terms of evaluation objectives and evaluation objects of different indicators as well as the selected indicators. For instance, when assessing whether a city is sustainable in terms of ecological performance, a comprehensive selection will be made from the above indicators, to ultimately obtain the eco-city indicator system through the scientific screening and weighting.

In the planning and design evaluation of the Shanghai World Expo, the research team absorbed and formed a series of indicators of interconnected ecological elements that were involved in the entire process of design, construction, operation and management of eco-city. The indicators in the system are mainly linked with environmental and ecological indicators, and can be classified as energy indicators, indicators of water resources, indicators of solid wastes, atmospheric environmental indicators, land indicators, and biological indicators according to natural factors (Wu 2009).

2.3 Structure of the City Indicator System: Parallel Mode and Progressive Mode

From the view of the internal structure of evaluation indicator system, there are mainly two ways of organization: parallel mode and progressive mode, between which the parallel mode is more common.

1. Parallel mode

The selected city indicators are usually divided into several parallel groups according to the evaluation theme, and the indicators in each group are calculated separately. Each group has one specific theme, and multiple groups constitute an entire system.

For example, the United Nations urban indicator system is organized in a "1 + 6 + 1" structure. The first "1" stands for background data, such as land use, male to female population ratio, urban per capita output, etc. These indicators will serve as the underlying data source for the subsequent evaluation, analysis and compound of six modules. And "6" refers to the six modules that constitute the evaluation indicator system, namely, social and economic development, infrastructure, transportation, environmental management, local government and housing. The last "1" represents a special indicator for more in-depth discussions of urban development and reflection of urban issues based on six modules (Wu and Ge 2001). It can be seen that the six core modules correspond to the six aspects of the evaluation indicator system, and the selection of indicators is based on these six aspects. The entire system operates with a parallel mode.

Another example is the construction of "Smart Nanjing". One of the important works is to build the evaluation indicator system of "Smart Nanjing". This system consists of four aspects of the indicator groups in parallel mode: infrastructure, urban wisdom industry, urban wisdom service, and urban wisdom humanities (Cheng et al. 2011). Such a system design not only helps to obtain data from each sub-item, but also improves the measures of urban practice according to the evaluation results of each sub-item.

2. Progressive mode

Progressive mode is another structure of city evaluation indicator system, which is relatively rarely used. The logic of the progressive mode is similar to Maslow's theory of demand, which classifies human needs (physiological needs, security needs, social needs, respect for needs, self-fulfilling requirements) from low to high (Milgram et al. 1999). The progressive mode of evaluation indicator system is also classified according to the importance expressed and transmitted by indicators. It is quite common to take basic indicators such as social and economic development data as the basic evaluation indicators to gradually focus on the concerned issues. Accordingly, the indicators involved will be more focused on a specific direction.

In the progressive mode, the grouping of indicators can usually be artificially set, and the progressive relationship of indicators can be naturally formed through methods such as scientific computing.

2.4   Indicator Calculation: Calculation of Simplex Indicators

If it is necessary to obtain the results of indicators, certain rules must be applied. The indicator calculation is the calculation rule that should be used to obtain a specific indicator. The indicator calculation is the most basic way to calculate the urban evaluation indicator system.

Professor Eugenio Morello of Milan Polytechnic University proposed a system that includes 146 indicators for urban sustainability assessment. These indicators fall into five categories: accessibility, society, environment, energy, and urban morphology. Each indicator contains the required basic information, calculation formulas, units, and reference ranges. For instance, in the level of accessibility, an indicator value called "Connections" is obtained by calculating the ratio of the number of streets connected and the number of intersections. And the formula involved is:

$$C = e/3(n - 2)$$

where C refers to connections, e refers to the number of connections, and n refers to the number of nodes. A separate calculation for these 146 indicators will ultimately reflect the sustainability of the city.

In addition to indicators that need to be obtained through calculation, data can also be obtained by means of statistics, sensing and other methods. These methods also include the transformation of qualitative indicators such as "yes" and "no" formed by judgement.

2.5   System Calculation: Indicator Weighing

The weight of the indicator usually reflects the relative importance of the indicator in the evaluation system (Su 2014). In the urban evaluation index system, it is very common to make the evaluation more fair and objective by giving the indicator or the indicator group weight.

There are many ways to determine the weight of indicators, which can be divided into two categories: subjective assignment method and objective assignment method, including many specific methods, such as addition integration method and analytic hierarchy process. In the subjective assignment method, the most important part is expert's judgement on the proportion of the reference results of indicators, which ultimately determines the specific weight percentage. For instance, in the application of urban tourism competitiveness evaluation, Su et al. (2003) used this process to evaluate. Another example: the analytic hierarchy process also has a wide range of applications. Multiple indicators are arranged and compared in pairs, and the reference weight obtained will be judged by experts to determine the final weight percentage (Peng et al. 2004).

3   Characteristics of City Indicator System

3.1   Scientificity and Practicability

The scientificity of the indicators should be understood as follows: the design of indicators needs to be based on scientific theory. Indicators must have a certain scientific connotation, with a clear purpose and a precise definition, and shall be measurable, easy to quantify and able to reflect the characteristics.

In the establishment of urban evaluation indicator system, it is required to be comprehensive and systematic. But when it comes to the evaluation of a specific issue, it is required to be simple and practical. Practicability means that the indicator system shall not only provide real support and guidance for policy makers, but also describe the city's social environment, stress and social responses, and be related to existing policy objectives.

3.2   Typicality and Comparability

City is composed of multiple systems. The relationship between systems and their behaviors are quite complex. Different evaluation objects have many optional indicators. In order to facilitate the description and explanation of issues, the most typical and representative indicators should be selected.

The urban evaluation indicator system should also conform to the principle of time and space comparability. As the differences between cities is an objective reality, it is better to use strongly comparable relative amount indicators and comparable indicators with common characteristics to form a relatively basic and unified system to measure. In this way, not only development of different cities can be compared, but it is also possible to be compatible with international research and convergence.

3.3 Dynamic and Operability

In a broad sense, the city is a dynamic set containing the development and changes of the natural, economic and social elements.[2] All elements go through waxing and waning promoting the city to move forward. In the design of the indicator system, changes of some indicators can be relatively stable or remain a stable position in the system within a short period of time, however, in the long term, we must carefully analyze the development of the system and update some outdated indicators.

The city evaluation indicator system not only has important theoretical research value, but should also have a practical value. When selecting an indicator, the selected indicator must be operable. In practice, indicators can be deleted and updated according to demands, or some of the original indicators can be integrated and subdivided to generate some necessary derivative indicators.

3.4 Prospective and Orienting

Comprehensive evaluation using the indicator system should not only reflect the current situation of the city, but also be forward-looking and indicate the future direction of the city by expressing the relationship between each element, such as the past and current resources, economy, society and the environment.

In designing the urban evaluation indicator system, it is also necessary to set aside a plan for the follow-up indicator system to update and upgrade. The orientation of the indicator system also needs to be reflected by the continuous updating of the indicator system itself.

---

[2]The definition of eco-city has not been defined yet, but according to the study of eco-city, it is not only the ecology of the environment, but also ecology of the society and the economy. The concept of ecology gradually converges to the concept of sustainable development.

3.5   *Hierarchy and Quantification*

Based on the amount of information, indicators can be divided into three levels: system, indicator and variable. It is shown that with the use of different objects, the total amount of information presents a shape of pyramid successively, but the concentration of information is ascending. This means that when designing the indicator system in practice, we should try to meet the principle of hierarchy. Otherwise, in the process of future implementation, some policies may be unable to be implemented, and the other policies may cause the implementation object difficult to be accepted.

At the same time, indicators as the system structure and behavior representation of urban sustainable development should be quantified.

4   Traps in the Structuring City Indicator Systems

In the construction of urban evaluation indicator system, the principle to choose a system design that is simple and easy to understand should be followed, so that the evaluation indicator system will be objective and practically operable. However, the construction of many urban evaluation indicator systems has the following problems.

(1) The target is unclear. Without a clear target, the evaluation is full of blindness.
(2) The system is redundant. There are overlapping, inclusion and causality relations between indicators.
(3) System imbalance. Too much emphasis is put on a particular aspect and others are weakened (especially in parallel mode).
(4) Indicators are invalid. some of the indicators selected cannot be obtained or lose timeliness.
(5) Indicators are vague. The indicators used for evaluation are difficult to measure or ambiguous.
(6) No prospectiveness. Timeliness is not fully taken into account in system design and selection of indicators.

In short, the city evaluation indicator system needs to be carefully designed by determining the key elements that need to be measured according to the evaluation targets, designing the overall structure of system, considering the indicators involved, taking full account of the availability and timeliness of indicators, choosing a reasonable weighting method to scientifically determine the weight percentage. Finally, after the completion of the indicator construction, trial operation should be carried out to ensure that the evaluation indicator system has no technical errors. For the above traps of construction, they shall be identified and avoided in the future design of city evaluation indicator system.

References

Chen M, Wang QC, Zhang XH et al (2011) The research on "intelligent city" evaluation indicator system—take Nanjing as an example. City Dev Res 5:84–89

Fu BR (2003) Design and application of city informatization measurement indicator system. Inf Sci 21(3):230–231

Guan YZ, Zheng JH, Zhuang SX et al (2007) Research on ecological civilization indicator system. China Dev 7(2):21–27

Milgram L, Spector A, Treger M (1999) Managing smart. Gulf Professional Publishing, Boston, p 308

Peng GF, Li SC, Sheng MK et al (2004) Research on confirming government performance measurement index weight through application level analysis. China Soft Sci 6:136–139

Su XP (2014) Residents' happiness indicator system research. Sci Technol Inf 9:85–86

Su WZ, Yang YB, Gu ZL et al (2003) City tourism competitiveness preliminary evaluation. Tourism Tribune 18(3):39–42

Tan YZ, Wu CF, Ye ZX et al (2003) City land sustainable utilization evaluation indicator system and method. China Soft Sci 3:139–143

Wen ZG, Li L (2007) Environmentally friendly city indicator system and its benchmarking management. Environ Prot 22:26–28

Weng R (2011) Law-ruled city indicator system research—taking Jiangsu as example. Southeast University, Nanjing

Wu ZQ (2009) The Shanghai world expo sustainable planning and design. China Architecture & Building Press, Beijing

Wu YY, Ge ZM (2001) UN city indicator system overview and evaluation. City Prob 3:13–15

Xu GF, Zhang Y (2011) City energy evaluation indicator system construction under the concept of sustainable development—taking Beijing as example. Resour Ind 13(5):5–9

Chapter 5
Intelligent City Evaluation Indicator Systems in China

Abstract At present, China does not have a unified intelligent city evaluation indicator system at the national level, but the government departments, enterprises and institutions are actively building it. This chapter provides an introduction of seven existing intelligent city evaluation systems in China, which are Pilot Intelligent City (District/Town) Certificated by Ministry of Housing and Urban-Rural Development of China (2012), Smart City Evaluation Indicator System by Ministry of Industry and Information Technology System (2012), GMTECH Smart City Development Level Evaluation (2016), China Wisdom Engineering Association Indicator Intelligent City (Town) Development Index (2011), Pudong Intelligent City Evaluation Indicator System (2012), Intelligent Nanjing Evaluation Indicator System (2011) and Ningbo Smart City Development Evaluation (2012).

Keywords Intelligent city · Evaluation indicator system · China

The effect of constructing intelligent city is not determined by a few people out of thin air. It should be based on a clear standard. Around 2010, China gradually launched a research on intelligent city evaluation indicator system, mainly under the guidance and supervision of the government, led by institutions and enterprises related to information technology industry based on their industrial advantages, technical conditions and practical experience after in-depth research and exploration.

At present, China does not have a unified intelligent city evaluation indicator system at the national level, but the government departments, enterprises and institutions are actively building it. Indication systems such as Pilot Intelligent City (District/Town) Certificated by Ministry of Housing and Urban-Rural Development of China (2012), Smart City Evaluation Indicator System by Ministry of Industry and Information Technology System (2012), GMTECH Smart City Development Level Evaluation (2016), China Wisdom Engineering Association Indicator Intelligent City (Town) Development Index (2011), Pudong Intelligent City Evaluation Indicator System (2012), Intelligent Nanjing Evaluation Indicator System (2011) and Ningbo Smart City Development Evaluation (2012) were released in succession. Such indicator systems have different focuses and indicators, but can be regarded as the indicator systems that are of symbolic significance and research value in China's intelligent city evaluation index system.

1   Pilot Intelligent City (District/Town) Certificated by Ministry of Housing and Urban-Rural Development of China

On November 22, 2012, the office of Ministry of Housing and Urban-Rural Development issued the Notice on the Work of the National Pilot Smart City and published the Interim Measures for the Administration of National Smart Cities and Pilot Intelligent City (District/Town) Certificated by Ministry of Housing and Urban-Rural Development of China (on trial). Two months later, two batches of national smart city pilot lists were announced by the Ministry of Housing and Urban-Rural Development, which starts the pilot work of China's smart city. In the above notice, cities (district/town) that are going to apply for national pilot smart city are required to refer to Pilot Intelligent City (District/Town) Certificated by Ministry of Housing and Urban-Rural Development of China (on trial) in order to prepare development plans for smart city, and to prepare implementation plans by setting practical construction goals of national intelligent city based on local reality so as to establish a corresponding system of policy, organization and funding.

In preparation of the indicator system, the Ministry of Housing and Urban-Rural Development not only referred to a number of domestic and foreign indicator documents, such as "Green GDP Indicator System", "Smart GDP Indicator System", "Recycling Economy Evaluation Indicator System", "Study on Evaluation Indicator System and Method of Urban Construction", "Standard Indicators for National Ecological Garden City", but also referred to the planning standards of areas that started researches on intelligent city at an early stage, such as Shanghai Pudong New District, Nanjing, and Wuhan.

The indicator system is divided into three grades: 4 indicators in the first grade, including the security system and infrastructure, intellectual construction and livability, intelligent management and service, intelligent industry and economy; 11 indicators in the second grade, including security system, network infrastructure, public platform and database, urban construction management, urban function enhancement, government services, basic public services, special applications, industrial planning, industrial upgrading and the development of new industries (see Table 1); 57 indicators in the third grade, in which details are provided for each item. The whole indicator system basically covers the contents of industry, people's livelihood, society, environment and infrastructure construction, and focuses on the overall development of the city and the solution for practical problems and gives answers to questions that the government concerns most: what the smart city is, how to construct smart city, what the advantages of smart city are and where to raise funds.

Table 1 National intelligent city (district/town) pilot system

First grade indicator	Secondary indicators
Security system and infrastructure	Security system
	Network infrastructure
	Public platform and database
Intelligent construction and livability	Urban construction management
	Urban function enhancement
Intelligent management and services	Government services
	Basic public services
	Special applications
Intelligent industry and economy	Industrial planning
	Industrial upgrading
	Development of new industries

Pilot Intelligent City (District/Town) Certificated by Ministry of Housing and Urban-Rural Development of China (on trial) is an indicator system released before Notice on the Work of the National Pilot Smart City. As a programmatic document in the field of smart city construction, it has obvious function in guiding and regulating local practices.

In the construction of the system, the indicator system emphasizes the role of the government as the main driving force of smart city, and pays attention to public goods and services that have a great influence, such as security system and infrastructure, construction and livability, industry and economy. The perspective of it shall not be limited to the promotion and use of intelligent technology; instead, it should pay more attention to the overall structure of social governance under the context of intelligent city construction.

Therefore, strictly speaking, the system is not an indicator system. The three grades of indicators (e.g. "development and implementation plan for smart city", "organizations" and "policy and regulations") are too broad, which cannot be quantified and are unable to point to micro operation. The guiding significance and symbolic significance of the system at the macro level is greater than the practical significance.

2   Smart City Evaluation Indicator System by Ministry of Industry and Information Technology System

As early as in 2002, the Ministry of Industry and Information Technology System had launched Smart City Evaluation Indicator System (on trial). This system was more focused on the indicator system for hardware such as infrastructure (Sun et al. 2013).

In June 2012, it is clearly proposed for the first time to guide and promote the healthy development of smart city in Opinions of the State Council on Vigorously Promoting the Development of Informatization and Safeguarding Information Security [Guo Fa (2012) No. 23]. Therefore, in order to know the situations for construction and management of smart city and to ensure that the work of smart city is conducted successfully, the informatization promotion division of the Ministry of Industry and Information Technology entrusted Computer and Microelectronics Development Research Center to prepare the Smart City Evaluation Indicator System (draft) and issued the Notice for Seeking Opinions on the Evaluation System of Intelligent City [Gong Xin letter (2012) No. 21]. Under the guidance of the Ministry of Industry and Information Technology, China Software Testing Center (2012) organized an event to seek opinions on evaluation indicators. Together with leading enterprises, research institutes, standardization organizations in the field of communications and intelligent computing, they founded China Smart City Industry Alliance (China Smart City Work Committee) in October 2013. Typical cities, mainstream ICT enterprise, authority experts in city management and information technology were invited to join the committee, and members of the committee and experts in their areas of expertise were encouraged to elaborate on the refinement of the indicators and the preparation for the construction guidance. To some extent, the evaluation indicator system of smart city proposed by the Ministry of Industry and Information Technology has a clear technical direction, and can be used as a detailed reference standard for enterprises and governments to build smart city.

On January 11, 2013, the annual meeting of China's smart city was held in Beijing. At the meeting, the China Smart City Evaluation Index System was released and the opening ceremony of the alliance was held. The indicator system released was prepared by Computer and Microelectronics Development Research Center (China Software Evaluation Center) after integrating the proposals and opinions of leaders and experts from domestic well-known IT enterprises and other relevant organizations, such as China Telecom, China Unicom, IBM, Microsoft, Huawei, Tai Ji, iSoftStone, Capinfo, Neusoft, Hua Di, Tsinghua Tongfang, Founder, Datang Telecom, China Intelligent Transportation, China Standard Software, Sinonet, Guangzhou Jiesai, H3C, SuperMap, Dalian Huaxin, China Electronics Society and China Software Evaluation Center. At final, they proposed 3 first grade indicators: smart preparation, smart management and smart service, including 9 secondary indicators: network environment, technical preparation, city operation management capability, construction process control, operation management mode and intelligent service coverage, as well as 45 inspection sites (see Table 2).

This evaluation system was based on the model of SMART theory. The model of SMART theory includes Service, Management, Application Platform, Resource and Technology (See Fig. 1), which covers most areas of smart city construction. From the above 5 parts, Resource and Technology are the basis; Application Platform is the most direct output (output layer); Management means to utilize intelligent management method to help with the planning and construction as well

Table 2 Smart city evaluation indicator system by ministry of industry and information technology system

General indicators	First grade indicator	Secondary indicators
Degree of smartness	Smart preparation	Network environment
		Technical preparation
		Security conditions
	Smart management	City operation management capability
		Construction process control
		Operation management mode
	Intelligent service	Intelligent service coverage
		Accessibility
		Processing efficiency

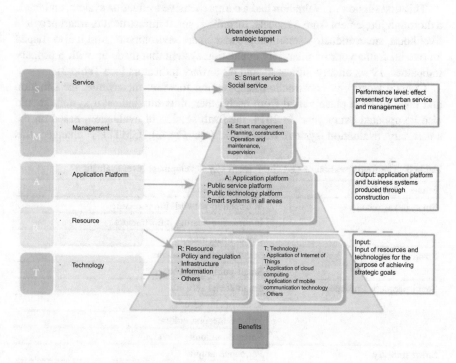

Fig. 1 SMART theory model

as maintenance in smart city. Service refers to all kinds of public service to the general public. Service and Management make up the performance layer.

It can be observed that, the Industry and information department version of smart city evaluation indicator system is focused more on the revolutionary change to the work style and efficiency which was brought about by the intelligent technology that based on the web of things. Among the 45 inspection points, there are plenty of quantitative indexes, such as optical fiber penetration ratio, Internet penetration

ratio, and response time for online service; meanwhile, there are also some qualitative indexes. From the above part, it can be found that the whole indicator system
is actually a comprehensive estimation to the present city's 'smart' operation.

3   Smart City Development Level Evaluation Sponsored by GMTECT

In Mar. 2011, the first session of smart city development level has been launched by
the Beijing GMTECT consulting company, LTD. On Sept. 9 the same year, the
First Session of Chinese Smart Cities development level evaluation report has been
officially released in Beijing, China, and revealed the result of this evaluation.

This first session of evaluation had a comprehensive evaluation system, and made
a thorough judgement from 6 aspects, including: smart infrastructure, smart people's
livelihood, smart industry, smart people and smart environment. And it also shaped
an overall framework of smart city evaluation system that made up with 6 primary
indicators, 19 secondary indicators and 42 tertiary indicators (see Table 3).

China's smart city construction keeps going forward, meanwhile, the situation
and level of each place varied from each other, thus this indicator system should
also be updated every year. In 2014, the forth session of evaluation, based on the
smart city evaluation system framework (see Fig. 2), GMTECT changed this

Table 3 The fourth session of GMTECT smart city development level evaluation

Primary indicator	Secondary indicators
Smart infrastructure	Information network infrastructure
	Information sharing infrastructure
	Urban infrastructure
Smart governance	Smart government affairs
	Smart public governance
Smart people's livelihood	Smart social security
	Smart health security
	Smart education culture
	Smart community services
Smart industry	Per capita output
	Input-output ratio
	Resource consumption per 10,000 GDP
	Integrate IT application with industrialization
Smart people	Information utilizing ability
	Innovation ability
	Quality of talents
Smart environment	Ecological protection
	Resource utilization
	Soft environmental construction

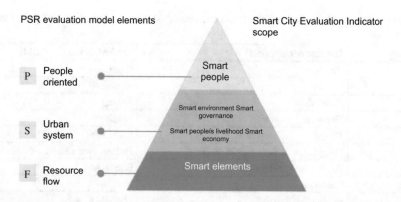

Fig. 2 PSF smart city evaluation system framework (*Source of the images* http://news.im2m.com. cn/375/16451684272.shtml)

Table 4 The fourth session of GMTECT smart city development level evaluation

Primary indicator	Weight	Secondary indicator	Weight
Smart infrastructure	25	Broadband network	10
		Foundation database completeness	5
		Urban cloud platform application	10
Smart management	20	Government collaborative work level	5
		Implementation of package solution for the industry	10
		The degree of social involvement to the public management	5
Smart service	20	The ability to integrate the services for the people's livelihood	10
		Government data open service	10
Smart economy	15	The per capita of patent numbers	5
		Energy consumption per 10,000 GDP	5
		The weight of added value of the information industry to the GDP	5
Smart people	10	User proportion of 3G/4G	5
		Per capita electronic e-commerce consumption	5
Security system	10	Development plan formulate	5
		Institutional framework and performance assessment	5
Total	100		100
Plus	5	Smart city pilot construction and application innovation, relevant honor and big events, etc.	5

indicator system into 6 primary indicators, 15 secondary indicators and a plus item, and gave weight to each indicator (see Table 4).

It can be found that, different from the standards of smart city that released by the government, GMTECt's report emphasizes on the involvement of common people,

Table 5 Results of the first and the fourth sessions of GMTECT smart city development level evaluation

Rank	The first session (2011)	The fourth session (2014)
1	Ningbo	Wuxi
2	Foshan	Shanghai
3	Guangzhou	Beijing
4	Shanghai	Ningbo
5	Yangzhou	Shenzhen
6	Hangzhouwan new district in Ningbo	Pudong new district
7	Pudong new district	Guangzhou
8	Shenzhen	Nanjing
9	Beijing	Hangzhou
10	Nanjing	Qingdao

and values the intelligent level and innovation ability of people. This demonstrates that GMTECT has a thorough comprehension on the connotation of smart city.

Meanwhile, based on this evaluation standard, GMTECT gave a quantification about the construction situation of domestic smart cities, and divided them into three levels according to their development stage: A (leader), B (chaser), C (preparer), and made a comprehensive evaluation to the level of smart city construction for each of them.

In the fourth session evaluation, GMTECT choose 100 domestic cities and gave a comprehensive evaluation to each city. Among them are 34 municipalities directly under the central government, provincial capital cities and cities at or above the provincial level, 57 prefecture-level cities, and 9 county-level cities. In 2014, the full score of smart city evaluation is 105, and the average score is 40.1. Wuxi has won the highest score (77.2), while Qiqihar has got the lowest score (17.6), with a gap of 59.6 points. The top 10 cities in the first and fourth sessions are showed in Table 5, which proves that the leading cities and regions remain, the same. Details of the top 10 cities see Table 6.

4   China Wisdom Engineering Association Indicator Intelligent City (Town) Development Index

On August 28, 2011, China Wisdom Engineering Association launched China Wisdom Engineering Association Indicator Intelligent City (Town) Development Index, putting forward the wisdom of the city development index system of the three aspects—happiness index, management index and social responsibility Index of wisdom city—as the three indicators. Employment income, medical treatment and public health, social security and other 22 items were taken as a secondary indicator. The information and network level, and other 86 items were taken as a

Table 6 The detailed information of Top 10 cities in the fourth session of GMTECT smart city development level evaluation

Rank	City	Smart Infrastructure	Smart management	Smart service	Smart economy	Smart people	Security system	Plus item	Total scores
1	Wuxi	20.4	14.5	13.5	9.5	7.8	8.5	3.0	77.2
2	Shanghai	19.4	12.5	15.5	7.5	7.2	7	5.0	74.1
3	Beijing	18.3	13	15	8	6.5	8	5.0	73.8
4	Ningbo	16.2	12	10	9	7.6	9	5.0	68.8
5	Shenzhen	19.2	10	10	8	9.7	7	4.5	68.4
6	Pudong new district	20.8	12.5	9.5	10.5	7.3	6	1.5	68.1
7	Guangzhou	17.4	14	12	6	7.4	5.5	3.0	65.3
8	Nanjing	17.4	11	11.5	6.5	7.8	7	3.0	64.2
8	Hangzhou	20.0	11	7	7.5	8.2	7.5	3.0	64.2
10	Qingdao	15.6	12	11	5.5	6.6	7	4.5	62.2

Table 7 China wisdom engineering association indicator intelligent city (town) development index

Primary indicators	Secondary indicators
Happiness index of intelligent city	Employment income
	Education
	Medical treatment and public health
	Social security
	Housing and consuming
	City cohesion
	Public service
	Organization and infrastructure construction
	Social service
Management index of intelligent city	Economic foundation
	Science and Technology Innovation level
	Human resource
	Human settlement
	Green-acting
	Ecological environment
Social responsibility index of intelligent city	Administration level
	Regional influence
	Communication capabilities for image
	Management and decision
	Duties for public affairs
	Rights and responsibilities
	Fiduciary duties

three-level indicator, and the community psychological assistance, volunteer culture and so on as a four-level indicator (see Table 7).

China Wisdom Engineering Association is a national comprehensive class association which was established by Mr. Qian Xuesen, initiated by Chinese Academy of Sciences, Chinese Academy of Social Sciences, Chinese Academy of Engineering, Tsinghua University, Peking University and other 20 scientific research institutions and famous universities, and in the charge of the Ministry of Education. This association along with other scientific research institutions and some authoritative experts from universities, after a year of preparation, investigated and researched some developing intelligent cities such as Shanghai Pudong New Area, Beijing, Wuhan, Zhejiang, Nanjing, Jiangsu, and Shaanxi and has consulted more than 100 experts, with reference of intelligent city's practical experience from the US, the EU, Japan, Singapore and other developed countries and regions,

Considering our national conditions and national 'The Twelfth Five-Year Plan', this set of intelligent city construction indicator system was put forward. The development index is a relatively complete intelligent city evaluation system, but because it focuses on the integrity of the system and pay less attention on the pertinence, which makes the indicator system too large, so the influence is not that huge.

5 Pudong Intelligent City Indicator System

On July 1, 2011, Shanghai Pudong Intelligent City Development Research Institute officially released "Intelligent City Indicator System 1.0", which was the first public release of the Chinese version of the intelligent city evaluation indicator system. On December 19, 2012, "Intelligent City Indicator System 2.0" was released.

Version 1.0 contains five dimensions (i.e., primary indicators) of infrastructure, public management and service, information services and economic development, humanities literacy and civic subjective perception, with 19 secondary indicators and 64 third indicators (see Table 8), and for each indicator a corresponding reference value is provided.

From start to the official releasing the Intelligent City Indicator System, it was a year or so. It mainly completed the following three aspects: a wealth of experience, a solid empirical basis and many expert seminars. It mainly considered the city 'intelligent' development as a concept and took the level of urban informatization, comprehensive competitiveness, green and low-carbon and humanities and other aspects of science and technology as an overall consideration.

In order to meet the needs of urban construction, Intelligent City Indicator System needs to be constantly innovative and continuously improved. After Shanghai Pudong Intelligent City Development Research Institute released the

Table 8 Pudong intelligent city indicator system 1.0

Primary indicator	Secondary indicator
Infrastructure	Level of broadband network coverage
	Level of broadband network access
	Level of investment and construction for infrastructure
Public management and service	Intelligent government service
	Intelligent mobility management
	Intelligent medical system
	Intelligent environment protection system
	Intelligent energy resource management
	Intelligent urban security
	Intelligent education system
	Intelligent community management
Information services and economic development	Level of industrial development
	Level of informationalized operation in companies
Humanities literacy and civic subjective perception	Level of citizen's income
	Civic culture and scientific literacy
	Level of public informationalized training
	Level of web-based civic life
Citizens' subjective perception	A sense of convenience of life
	A sense of secure of life

Indicator System 1.0, it continued to further improve the indicator system with collaborative research. After about one year of continuous research, "Intelligent City Indicator System 1.0" has been more substantially improved. In December 2012, China (Shanghai) Intelligent City Summit Forum, "Intelligent City Indicator System 2.0" was released. It was a latest masterpiece which was accomplished through empirical research, repeated demonstration and modification.

This indicator system is based on the concept of "intelligent" city, with the level of urban informatization, comprehensive competitiveness, green and low-carbon, humanistic science and technology and other factors fully considered. It is divided into 6 dimensions as infrastructure, public management and service, and other dimensions, including 18 secondary indicators, 37 tertiary indicators (see Table 9). Compared with version 1.0, version 2.0 added one more indicator which is "soft environment construction for Intelligent City". It also cut down some secondary indicators and tertiary indicators.

Table 9 Pudong intelligent city indicator system 2.0

Primary indicator	Secondary indicator	Tertiary indicator
Infrastructure	Level of broadband network construction	Domestic fiber access rate
		Major public places WLAN coverage
		Average network access level
Public management and services	Intelligent government services	Administrative examination and approval matters online processing level
		Online circulation rate of government non-secret documents
	Intelligent mobility management	Electronization rate of bus stops
		Compliance rate of citizen traffic guidance information
	Intelligent medical system	Inputting rate of public fitness electronic records
		Usage rate of electronic medical record
	Intelligent environmental protection	Automate monitoring ratio of environmental quality
		Level of major pollution monitoring
	Intelligent energy and resource management	Installation rate of domestic smart meters
		The proportion of new energy vehicles
		Digital energy-saving ratio of buildings
	Intelligent city security	Establish rate of emergency response system for major emergencies
		Monitoring level of dangerous chemicals transportation
	Intelligent education system	Urban education expenditure level
		The proportion of network teaching
	Intelligent community management	Capability of Community Integrated Information Service

(continued)

Table 9 (continued)

Primary indicator	Secondary indicator	Tertiary indicator
Information service and economic development	Industrial development level	The added value of information services that accounted for the proportion of GDP
		The proportion of employees that in the information service industry to the total number of employees in the society
	Enterprise informationalized operation level	Website building rate of enterprises
		Enterprise e-commerce behavior rate
		Enterprise informationalized system utilization rate
Humanities and scientific literacy	Income level of citizens	Average disposable income
	Civic culture and scientific literacy	College and above accounted for the proportion of the total population
	Level of networked public life	Online access rate
		Domestic online shopping rate
Citizens' subjective perception	Sense of convenience of life	Access to transit information
		Extent of convenience for medical treatment
		Extent of convenience for government services
	Sense of safety in daily life	Satisfaction for food and drug safety electronic monitoring
		Satisfaction for environmental safety information monitoring
		Satisfaction for traffic safety information system
Soft environment construction for intelligent city	Planning and design for intelligent city	Development and planning for intelligent city
		Organization and leadership mechanism of intelligent city
	The creation of the atmosphere for intelligent city	Forum meetings and training levels of intelligent city

Pudong Intelligent City Indicator System and its years of informationized technology and intelligent city practice are inseparable and mutual confirmed. Pudong intelligent city practice is characterized by people's livelihood, so the establishment of its indicator system is also from the perspective of personal feelings, such as urban humanities literacy, public perception of the city, urban soft environment construction and so on.

In addition, in the selection of indicators, version 2.0 fully considered the collectability, representative and vertical and horizontal comparability of the three indicators, making the whole set of indicators closer to the civic feeling, while more easily to compare and measure at the same statement.

6   Nanjing Smart Nanjing Evaluation System

Deng (2010), who worked in Nanjing Information Center, based on the analysis of the evaluation indicator system of urban informatization, summarized the evaluation indicator system of intelligence in Nanjing according to the connotation and development characteristics of the intelligent city. The primary tier of the indicators includes the fields of urban network interconnection, intellectual industry, intellectual services and intellectual humanities, with a total number of 21 evaluation indicators, and analyzed these indicators one by one by using Nanjing data.

Based on Deng's theory, Chen et al. (2011) extended the field of intelligence and humanities from five items to seven. It finally formed four primary indicators as infrastructure, intelligent industry, intelligent services, and intelligent humanities, and a total number of 23 secondary indicators, without tertiary indicators (see Table 10). The evaluation indicator system of smart Nanjing is a kind of indicator system which is based on local practice in order to clarify the development direction and avoid the construction risk and guide the concrete operation.

Table 10 Smart Nanjing evaluation indicator system

Primary indicator	Secondary indicator
Infrastructure	Wifi Coverage rate
	Fiber access coverage rate
	Average network bandwidth
	Number of national key laboratory
	Smart grid technology and equipment applications
Smart industry	Investment amount in fixed assets of smart industry
	Expenditure on the R&D of the smart industry
	Proportion of smart industry to GDP
	Number of employees in the smart industry
	Total number of patent applications for invention of smart industry
	E-commerce transactions
	Electricity consumption per unit of GDP
Intelligent services	Government executive efficiency index
	Collaborative application system
	Application and popularity of intelligent public service
	Construction capital investment of intelligent public service
Wisdom of humanities	Urban labor productivity
	Proportion of college and above gradates to the total population
	Proportion of Information service industry practitioners to the total social practitioners
	General index of informationalized level
	Survey of urban public service satisfaction
	The proportion of cultural and creative industries to GDP
	Evaluation of international cultural and sports exchange activities

Nanjing is one of the cities that carried out the practice of intelligent city very early in China. One of the biggest features of this evaluation system is that the number of indicators is small and it is theoretically quantifiable. In addition, the evaluation system with particular emphasis on intelligent city construction in the humanities and cultural performance, echoed the Nanjing Information Center Director Mr. Tong Longjun in the "from Digital Nanjing to Smart Nanjing" as the theme of the speech mentioned:

> Talents are our most important resources, we talked about a lot of problems when building a smart city, but an aspect neglected is the wisdom of humanities. We help many cities to introduce talents, which is a kind of talent protection. In any case, no matter how the case changes, the aim never changed. An intelligent city ultimately relies on people to be achieved.

As a local intelligent city evaluation system, the Smart Nanjing evaluation indicator system is similar to Pudong Intelligent City Evaluation Indicator System. It pays more attention to the indicators of the collection, operation and quantification. While catering to reality, it is inevitable to respond and adjust according to the new development trend continuously.

7   Ningbo Intelligent City Development and Evaluation Indicator System

On September 3, 2011, The People's Government of Ningbo invited several relevant domestic and foreign scholars and relevant leaders to the Expert Consultation Meeting of Intelligent City Development and Evaluation Indicator System in Ningbo Shangri-La Hotel. Ningbo Intelligent City Development Evaluation Indicator System was researched and drafted by Ningbo Academy of Smart City Development, which is one of the main declaration (and in charged) institutes of the 2011 Ministry-City cooperation of a major national software science research project "Key Issues of Intelligent City", and some research teams from Zhejiang University and other universities, and some other consulting institute. This indicator system consists of six primary indicators, 19 secondary indicators, and 39 tertiary indicators. The specific assessment key points are 119 items in all. Among them, six primary indicators are smart people, smart infrastructure, smart governance, smart living, smart economy and smart environment. When choosing the specific indicators, the indicator system fully considers the people's basic necessities and the feelings of happiness. There are many measures that are closely related to people's livelihood, that people can "see it, feel it". It is a set of measurement system that based on personal feelings (Quan and Chen 2011).

In 2012, Gu and Qiao (2012) deepened this evaluation system, increased the level of "Smart Planning and Construction", and adjusted the secondary and tertiary indicators. Finally, a set of intelligence city evaluation indicator system was built, which was based on objective indicators, and has a strong operability, including seven primary indicators, 21 secondary indicators, and 48 tertiary indicators (see

Table 11 Ningbo intelligent city development and evaluation indicator system

Primary indicator	Secondary indicator
Smart people	Human resource
	life-long learning
	Information consumption
Intelligent infrastructure	Communication facilities
	Information sharing infrastructure
Smart governance	E-government
	Public participation of government decisions
	Public service investment
Smart living	Social insurance
	Medical treatment
	Mobility
Smart economy	Economic strength
	Smart industry
	R&D capabilities
	Output consumption
	Industrial structure and contribution
Smart environment	Waste disposal capacity
	Environmental attraction
Smart planning and construction	Urban and rural integration
	Space layout
	Smart building

Table 11). Among the tertiary indicators, there are 44 objective indicators, accounting for 91.7%; 34 data easy-accessed indicators, which accounting for 71%, indicating that the indicator system has a strong operability and applicability.

Ningbo is also one of the leading areas of intelligent city construction in the eastern coastal areas of China. Since September 2010, after making the decision to build an intelligent city, Ningbo has won the "China City Information Technology Outstanding Achievement Award", "Smart Zhejiang" honors. Ningbo has established the intelligent city planning and development institute, and united the United Nations Ministries, local governments, domestic scientific research institutions and famous universities to undertake research tasks. Beside focusing on infrastructure, people's livelihood, industry, environment and other major content, Ningbo intelligent city development evaluation indicator system will also be an evaluation object of "intelligent planning and construction", focusing on urban and rural integration, spatial layout and smart buildings and other aspects of progress.

Earlier, as a part of the evaluation indicator system of "Key Issues in the Study of China Intelligent City", and also conducted a comprehensive assessment of the districts of Ningbo, the score indicated that the intelligent degree differences were obvious. Intelligent city construction also had certain relevance with each county's economic development.

8 Conclusion

Viewing from the main body of the preparation and distribution of the indicator system, there have been seven existing intelligent city evaluation systems issued by government departments or research institutions with government background. The compilation process is led by experts in the field and the relevant scientific research institutes and enterprises Institutions. The focus, authority and influence of the indicators are affected by the preparation of the main body to varying degrees.

Ministry of Industry, and Ministry of Housing and Urban-Rural Development of PRC China, as the national competent ministries, put more emphasis in their preparation of the indicator system on the integrity of the system and qualitative analysis, not as local standards (such as Nanjing, Ningbo, etc.) focus on the collection and quantification of the indicators. Among them, the Ministry of Industry, involved in information technology earlier, is more closely linked with the industry, for which its indicator system is more concerned about the effectiveness of intelligent technology presented. Ministry of Housing and Urban-Rural Development of PRC China is more focused on the implementation of intelligent city space. For example, in the tertiary indicators of network environment, indicators from the Ministry of Industry are more detailed, including the average speed of the Internet, domestic fiber accessing rate, 3G network coverage, the use of 4 M and above broadband, Internet penetration, smart phone ownership rate, the proportion of mobile broadband users, while indicators from Ministry of Housing and Urban-Rural Development of PRC China are just wireless network, broadband network, the next generation of radio and television network. In addition, in the city management, the indicator system of Ministry of Housing and Urban-Rural Development of PRC China is much more abundant than the Ministry of Industry. Indicators such as urban and rural planning, construction market management, water supply systems, drainage systems, water-saving applications, gas systems, waste separation and treatment, heating systems, lighting systems, underground pipelines and space integrated management, intelligent logistics, intelligent payment, smart commerce and other indicators are not involved in indicators of the Ministry of Industry. In conclusion, the indicator system of the Ministry of Industry put more emphasis on the test of intelligent city of technical support capacity, but the indicator system of Ministry of Housing and Urban-Rural Development of PRC China put more emphasis on the embodiment of intelligent city specific urban planning and management aspects.

From the perspective of the content of the evaluation content, it basically concentrated in the hardware, environment, services and humanity four aspects, all built layer by layer. But with a scrutiny of each evaluation indicator system, there are still differences between each of them.

For example, the evaluation system issued the Ministry of Industry more focused on infrastructure and other hardware; evaluation indicator system published by Shanghai Pudong is the first country-specific evaluation system which is developed by the region, has a strong regional characteristics; China intelligent city (town) development indicator evaluation system is more inclined to public management and public policy from the perspective of the intelligent city as a whole analysis; the assessment report issued by Guomai company mainly for dividing the stages of China intelligent city development. It classifies some typical intelligent city among our country.

China intelligent city evaluation indicator system construction still needs to be further developed. First of all, the essence of intelligent city should be cleared. Based on this, relevant influential factors shall be put forward for its various components, so as to determine the evaluation criteria for each element, and implement specific evaluation indicators.

In the next step of building the evaluation indicator system, we should pay attention to the following four characteristics:

(1) Unity. The specific evaluation indicators of the elements should be as consistent as possible with the national statistical caliber in order to facilitate data collection, processing and application.
(2) Diversity. There are differences in the background, stage and problems of the development of local intelligent cities, and different factors and specific indicators should be set up to encourage the localized intelligent city construction.
(3) Operability. Pay attention to the collectability, quantization, and applicability of each vital influential factor. Ensure the evaluation indicator system can be with the Internet of Things, large data, and cloud computing and other intelligent city practices. With the specific input and output of the intelligent city construction, the indicator system must also be with the development of intelligent city and advanced with the times.
(4) Detail. Pat attention to the further refinement of the elements. For example, It can be implemented with measuring the number of sensors per unit area to assess the degree of intelligence of urban infrastructure, in order to make sure that the city intelligence degree is true and to avoid the choice of factors too macro Resulting in false results.

References

Chen M, Wang QC, Zhang XH et al (2011) The research on "intelligent city" evaluation indicator system—take Nanjing as an example. City Dev Res 5:84–89

China Software Testing Center (2012) Intelligent city evaluation indicator system is coming soon—22 well-known ICT industry enterprises discuss intelligent city evaluation indicator system. http://www.cstc.org.cn/templet/default/show_xwzx.jsp?article_id=122104&id=1380. Accessed 30 Oct 2014

Deng XF (2010) The research on "intelligent city" evaluation indicator system. Dev Res 12:111–116

Gu DD, Qiao W (2012) Research on the construction of intelligent city evaluation indicator system in China. Future Dev 35(10):79–83

Quan JY, Chen B (2011) Expert consultation discussion meeting on intelligent city development evaluation indicator system held in Ningbo. http://www.zjkjt.gov.cn/news/node01/detail0104/2011/0104_24563.htm. Accessed 9 Sep 2014

Sun J, Liu YT (2013) Intelligent city evaluation indicator system present condition analysis. Inf Const 2:30–31

Chapter 6
Intelligent City Evaluation Systems in West

Abstract This chapter investigates five released intelligent city evaluation systems in West, which are TU Wien Indicator System (2007), International Digital Corporation Indicator System (2011), Intelligent Community Forum Indicator System (1994), IBM Indicator System (2010) and Ericsson Indicator System (2010). Moreover, the differences between domestic and foreign indicator systems are analyzed from the perspectives of research leader, system construction and indicator selection. In sum, compared with the domestic government-led compilation of intelligent city evaluation indicator system, foreign research institutions (including universities) and enterprises are the ones to release the relevant evaluation indicator system; The construction of indicators system in foreign countries is more flexible, while the domestic indicator system is more alike in the large framework and structure; Compared with the domestic indicator system, the selections of indicators system in foreign countries has higher universality, and is more international.

Keywords Intelligent city · Evaluation indicator system · West

1 TU Wien Indicator System

In October 2007, Prof. Giffinger et al. (2007) from the Technische Universität Wien (TU Wien) published a report entitled "Ranking of the Medium European Intelligent Cities". It evaluated the intelligent city through six aspects, namely the smart economy, smart people, smart governance, smart living, smart mobility, and smart environment, and then updated the data and indicators, introducing 2.0 (2013) and 3.0 (2014) two versions.

Although the report of this evaluation system is aimed at the medium-sized cities in Europe, the six aspects have basically covered all the development key point of the intelligent city. The report focuses on the important issues of social development in the crowd, life, environment and other social issues. The proposed elements covered the spirit of innovation, entrepreneurial spirit, environmental protection,

© Springer Nature Singapore Pte Ltd. and Zhejiang University Press 2018
Z. Wu, *Intelligent City Evaluation System*, Strategic Research
on Construction and Promotion of China's Intelligent Cities,
https://doi.org/10.1007/978-981-10-5939-1_6

resource management, social cohesion, racial diversity, lifelong learning participation and other soft indicators. The concept of intelligent city is not confined to the
ICT level. The significance of the report is the establishment of the EU-wide
cross-border application of the indicator system for the first time, ranking European
secondary cities according to the specific indicators. The results have important
reference value to China intelligent city evaluation.

1.1 Sample Selecting

According to certain requirements, the sample cities were selected: first requirement
is that the city must be a medium-sized city, and must have access to the database.
The samples are mainly from the database of European Spatial Planning
Observation Network (ESPON) which covered almost 1600 cities. Specific
screening conditions are as follows:

- The population must be between 100,000 and 500,000: to ensure that the city
  belongs to medium-sized cities;
- There is at least one university: excluding the low-level education of the city;
- The total urban population must be less than 1.5 million: excluding cities that
  belong to the metropolitan area;
- Cities included in the Planning for Energy Efficient Cities (PLEEC) project can
  be selected directly.

1.2 Indicator System

From the definition of the intelligent city, 6 Characteristics were summarized from
the views that academic community discussed about intelligent city, which are smart
economy, smart people, smart governance, smart mobility, smart environment, smart
living. Each Characteristic is subdivided into 3–7 factors. Each factor can be assessed
with a number of indicators in order to be compared and calculated through Data (see
Table 6.1). The score for the German city Erfurt is shown in Fig. 6.1.

1.3 Data Normalization

Because comparing the different indicators requires normalizing the data, the data
were grouped into the [0, 1] interval by z-score normalization formula to facilitate
further comparisons. To compare the different cities in the "smart economy", "smart
people" and other attributes of the distinction, it was necessary to start from the
sub-indicators.

Table 6.1 TU Wien indicator system

Property	Factor
Smart economy	Innovative spirit
	Entrepreneurship
	Economic image and trademarks
	Productivity
	Flexibility of labour market
	International embeddedness
Smart people	Level of qualification
	Affinity to life long learning
	Social and ethnic plurality
	Flexibility
	Creativity
	Cosmopolitanism/open-mindedness
	Participation in public life
Smart governance	Participation in decision-making
	Public and social services
	Transparent governance
Smart mobility	Local accessibility
	(Inter-)national accessibility
	Availability of ICT-infrastructure
	Sustainable, innovative and safe transport systems
Smart environment	Attractively of natural conditions
	Pollution
	Environmental protection
	Sustainable resource management
Smart living	Cultural facilities
	Health conditions
	Individual safety
	Housing quality
	Education facilities
	Touristic attractively
	Social cohesion

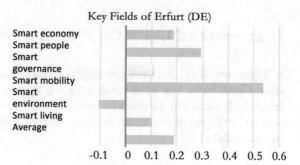

Fig. 6.1 Urban sub-index *Source* Official website of TU Wien indicator system http://www.smart-cities.eu

Fig. 6.2 Distribution of 77 sample cities that the TU Wien indicator system chose. *Source* Official website of TU Wien indicator system http://www.smart-cities.eu

The problem with data coverage is that the data of some indicators are incomplete and cannot cover all 77 sample cities (see Fig. 6.2), so the weights of the indicators should be adjusted, for instance, the indicators with data covering more than 70 cities should be given more weight than ones with data covering 60 cities. This reduces the impact of missing data on rankings, and of course the ideal situation is comprehensive data being obtained.

1.4   Ranking

Based on the above data and indicator system, 77 medium cities in Europe were listed (the top 20 in Table 6.2), including detailed indicators of each city.

2   International Digital Corporation Indicator System

In September 2011, International Digital Corporation (IDC) presented the IDC Smart Cities Index in a white book analyzing Spainsh intelligent city for the first time, which evaluated 44 cities in Spain. In the same year, it was applied in the German Intelligent city analysis. In 2012, the new rankings were introduced based on data updates.

Table 6.2 Ranking list of TU Wien indicator system (2014)

Country	City	Smart economy	Smart peoples	Smart governance	Smart mobility	Smart environment	Smart living	Overall ranking
Luxembourg	Luxembourg	1	18	56	4	16	4	1
Denmark	Aarhus	2	3	6	3	19	27	2
Sweden	Umeaa	24	5	2	34	1	13	3
Sweden	Eskilstuna	21	1	7	24	3	41	4
Denmark	Aalborg	10	11	5	14	14	10	5
Sweden	Joenkoeping	32	13	3	11	2	26	6
Denmark	Odense	13	9	4	20	9	40	7
Finland	Jyväskylä	23	8	1	47	5	25	8
Finland	Tampere	16	2	15	31	12	14	9
Austria	Salzburg	27	24	29	2	27	1	10
Finland	Turku	20	6	12	15	18	29	11
Finland	Oulu	14	4	9	39	13	35	12
Austria	Innsbruck	35	27	26	12	6	3	13
Austria	Linz	11	23	31	8	25	7	14
Slovenia	Ljubljana	6	7	34	33	21	21	15
Austria	Graz	26	21	33	9	28	2	16
Finland	Eindhoven	5	12	24	1	49	49	17
Germany	Regensburg	4	17	37	10	37	11	18
France	Montpellier	29	20	16	46	4	30	19
Belgium	Gent	15	29	27	6	41	9	20

Source Official website of TU Wien indicator system http://www.smart-cities.eu

Table 6.3 International digital corporation indicator system

Dimension	Unit
Smartness dimension	Government
	Buildings
	Mobility
	Energy and environment
	Services
Enabling force	Information and communication technologies
	People
	Economy

This indicator system examines the construction of the intelligent city from the aspects of the Smartness Dimension and the Enabling Force. 5 Smartness Building Blocks were extended in the Smartness Dimension, as government, architecture, transportation, energy and environment, services. Enabling force includes information and communication technology, citizens, and economy three units (see Table 6.3), then subdivided into 23 Evaluation Criteria and 94 lower level indicators.

Sample cities were selected in Spain. All the cities were ranked according to the population, and ones with population more than 150,000 were selected, which is totally 44 (see Fig. 6.3).

According to the indicators of each system mentioned, the data from each sample city (the proportion of Smartness Dimension and Enabling Force is 80% and 20% respectively) were analyzed. Here came the following results.

(1) 5 leading cities, Malaga, Barcelona, Santander, Madrid and San Sebastian are evenly developed and in a leading position in all dimensions.
(2) 10 contenders. These cities have relatively high scores in each unit of wisdom, but are not in a leading position.

These cities have the potential to be developed into leading cities. If their governments adopt the correct strategic line, they will make a breakthrough in government management, facilities construction and industrial development. If Zaragoza and Bilbao take positive action in the aspects such as intelligent government, intelligent building, and intelligent traffic, they will soon enter the rank of the leaders.

(3) 21 players. These cities are at an average level in all dimensions of the construction of intelligent city. In general, they have made a certain attempt in the intelligent city aspect, but still not fully promoted.
(4) Followers. These cities are in a backward position in various dimensions.

The analysis results are plotted as a matrix. As shown in Fig. 6.4, the horizontal axis is the enabling force of the city. It describes the government's enthusiasm and support for the construction of the intelligent city. The vertical axis is the Smartness Dimension, that is, the maturity of each smartness field.

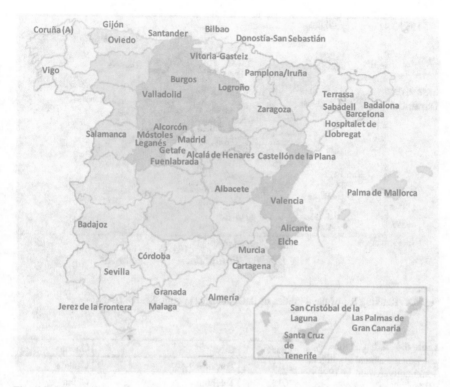

Fig. 6.3 Distribution of 44 sample cities of the indicator system. *Source* IDC

The indicator system stands in terms of technology, considering the development state of the technical application of the intelligent representation, and the economy population and technology development level and the development basis that it relied on as two different objects, which contains some certain merits, but It is inevitable that the secondary indicators will have intersection and overlap in the concept and directivity.

3   Intelligent Community Forum Indicator System

The Intelligent Community Forum (ICF) selects the Intelligent Community of the Year each year. The process is: a list of seven candidate cities in January each year is submitted, then a third party assessment agency analyzes the data of candidate cities, ICF organizers do the field research and submit research reports at the same time, and ultimately the government leaders, business leaders, academic pioneers and consulting firms determine the rank, and then announce the winner at the end of year. The cities of the annual Intelligent Community Award in 1999–2014 are shown in Table 6.4.

Fig. 6.4 Ranking list of the 44 cities of IDC indicator system. *Source* IDC

	Year	City	Country
Table 6.4 A list of the city that won the annual Intelligent Community Award in 1999–2014	Year	City	Country
	2014	Toronto	Canada
	2013	Taichung	China
	2012	Riverside	America
	2011	Eindhoven	Netherlands
	2010	Suwon	Korea
	2009	Stockholm	Sweden
	2008	Gangnam District, Seoul	Korea
	2007	Waterloo	Canada
	2006	Taipei	China
	2005	Mikata	Japan
	2004	Glasgow	America
	2002	Calgary	Canada
		Seoul	Korea
	2001	New York	America
	2000	LaGrange	America
	1999	Singapore	Singapore

Source http://www.intelligentcommunity.org

The selected cities don't need to apply the cutting-edge intelligent technology, but they must have paid much attention to social equity, knowledge innovation, sharing developing experiences to the whole world in the process of technology promotion with the routine technology.

Strictly speaking, the evaluation criteria of intelligent city established by American Intelligent Community Forums can never be an evaluation indicator system. These evaluation standards have five different aspects, including: broadband connectivity, knowledge workforce, innovation, digital inclusion, marketing and advocacy. Spread and application of ICT is the starting point, and it should be associated with more wide range of topics, like environment and society at the same time.

(1) Broadband Connectivity. New infrastructure can effectively manage the development of economics. Intelligent community has raised a better future established on the broadband connection.
(2) Knowledge Workforce. Knowledge is productivity. When they are collected, processed, and analyzed, they can not only create value, but also help factory production transform into laboratory R&D and site construction transform into office online.
(3) Innovation. Global economy is driven by innovation. As the internet is spreading, the difference is decreasing between different cities, and innovation will be the important indicator for competitiveness. The new, cheaper, and rapid technology brings change to medical treatment, agriculture, entertainment and education areas. It promoted innovation, and meanwhile raised standards for joining in the club of global economy. People need more knowledge to control the information flow.
(4) Digital Inclusion. Intelligence Community should eliminate information isolation, the difficulty of accessing to information media, which is caused by poverty, lack of skill, prejudice, or geographical reason, making some people in the dry tree due to a lack of information. Digital inclusion can eliminate the segregation by way of broadband access, skill training, and macro-willingness.
(5) Marketing and Advocacy. Facing with the challenge and competition from globalization, intelligent community needs to disseminate its progress in life, working, and commercial fields through powerful promotion.

4   IBM Indicator System

IBM is an active advocator of the intelligent earth, and also a solution provider for intelligent city. In order to serve the commercial development, IBM built its own intelligent city evaluation indicator system. In 2010, the IBM Business Value Institute proposed the following principles for assessing urban wisdom.

(1) Being tailored. Intelligent city assessment must consider each city with different visions and goal priorities. To meet this requirement, there is a way to use the weighted scoreboard method to conduct a comprehensive assessment of the object. The scoreboard should contain the relevant standards for each system,

define and evaluate the overall status and performance of each system, as well as the city as a whole, according to the importance of the city.

(2) Based on the overall urban view, a large number of urban systems mutually influence each other and the change of one system inevitably affects the others. Therefore, the assessment process should take into account the entire urban framework. For example, if a city independently evaluates a system (such as energy) without acknowledging its dependency on other systems (such as transportation, commerce, and water supply) and the impact on energy consumption, the results may mislead the city to take a so-called corrective action that harms the entire city.

(3) A comprehensive progress of the whole system should be fully measured. The intelligent city evaluation should have a comprehensive understanding of how each system is transformed when an intelligent solution is applied. This requires detailed criteria and variables of the necessary conditions, management of the system, usage of the solution, and the expected results for each system. By using well-designed standards, the city can fully understand the situation of each system transformation and the three major technical systems (Internet of Things, the Internet and intelligent technology) situation, for each application carrier to put forward a corresponding technical solution at different levels.

(4) Comparability and appropriate measurement with the basis of equivalent cities are required. It is equally important to measure what aspects to measure and how it is measured based. Cities can choose cities that have the same key characteristics, challenges, and priorities to compare with, in order to help officials of the city share their best practices and insights in various social activities in the future.

IBM's evaluation criteria and elements examples are shown in Table 6.5. The IBM indicator system is characterized by combination of the application carrier and the technology system, to identify the current technical gaps or weak links. Its purpose is to serve the IBM global commercial strategy of intelligent earth.

Therefore, IBM indicator system is more concerned about the needs of specific objects, and more emphasizing on understanding the extent to which the technical program can achieve.

5   Ericsson Indicator System

Like IBM, Ericsson is also an active company in the field of intelligent city business promotion. Ericsson indicator system is called the Networked Society City Index. It was released in 2010, updating the data of the indicator system and the city evaluation annually.

The core idea of the Ericsson indicator system is to distinguish ICT maturity from ICT's "Triple Bottom Line" (TBL). ICT development maturity includes three dimensions, namely infrastructure, affordability, and application, and the "triple

Table 6.5 Examples of IBM evaluation criteria and elements

System	Necessary conditions	Management	Intelligent system	Effectiveness
Urban services	Local government expenditure	Coordinated service delivery	E-government	The efficiency and effectiveness of public service delivery
	Local government personnel		The application and usage of ICT in service delivery and management	
Peoples	Investment on education, health, housing, public safety and social services	Joint management and coordination of service delivery	ICT and the intelligent technology in the application and use of the human and social services	Effectiveness of education, health, housing, public safety and social services
Commerce	Capital acquisition, administrative burden, trade barriers, commercial real estate	Joint and effective supervise and management of commercial systems	Usage of ICTs in enterprise; new intelligent business processes; intelligent technical field	Value-added, business creation, innovation
Mobility	Transport infrastructure and public transport investment; infrastructure quality	Joint supervision of mobility systems	The use of intelligent mobility technology; congestion charges	Congestion level; urban traffic conditions; energy consumption of mobility system
Communication	Investment in communication infrastructure	Coordinating supervision of communication systems	High speed broadband; Wi-Fi	Quality and accessibility of communication system
Water supply	Investment in water supply infrastructure; supply of clean water; discharge of sewage	Management and supervision of water supply systems	Intelligent technology in the field of water supply management	Water use; water waste/loss
Energy	Investment in energy infrastructure	Coordinating supervision of energy system	The construction of intelligent grid; the use of the intelligent meter	Energy waste/loss; reliability of energy supply; renewable energy

Source IBM

Table 6.6 Ericsson indicator system

ICT development maturity		TBL affection	
Infrastructure	Width quality	Society	Fitness
	Accessibility		Education
Affordability	Cost rate		Inclusive
	IP conversion costs	Economy	Efficiency
Application	Technology application		Competitiveness
	Personal application	Environment	Resource
	Public and market applications		Pollution
			Climate change

Refer to: http://www.ericsson.com/thinkingahead/networked_society/city-life/city-index/about

Table 6.7 Ericsson calculated the top 10 intelligent cities in 40 cities (2014)

Ranking	City
1	Stockholm
2	London
3	Paris
4	Singapore
5	Copenhagen
6	Helsinki
7	New York
8	Oslo
9	Hong Kong
10	Tokyo

bottom line effects" are broken down into three main aspects as society, economy and the environment (see Table 6.6).

By assigning variables, the Ericsson Indicator System 2014 version calculated the Network Index for 40 cities worldwide (see Table 6.7). The top three are Stockholm, London and Paris. Among Chinese cities, Hong Kong ranked 9th, Beijing ranked 26th, and Shanghai ranked 28th.

Ericsson works closely with the International Telecommunication Union (ITU), and its indicator system is more influential than the IBM indicator system. Ericsson has been using the so-called "Connected City" to replace the popular title "smart city". Its unique evaluation system is concise and clear, with a very clear purpose, that is, through the application of ICT to enhance a city in the economic, environmental and social sustainable development level. Although this formulation is not that new, but because of the support of quantitative data, it is still a kind of evaluation method that can easily get approval.

6 Conclusion

Compared with the domestic government-led compilation of intelligent city evaluation indicator system, foreign research institutions (including universities) and enterprises are the ones to release the relevant evaluation indicator system. From the perspective of the construction of the system, the enterprise evaluation indicator system is more focused on the analysis of the intelligent city from the technical angle, while the research institutions are standing in a more comprehensive and objective perspective to understand the intelligent city, so the evaluation indicator system that the latter one put forward is more persuasive.

In addition, the construction of indicators system in foreign countries is more flexible, while the domestic indicator system is more alike in the large framework and structure, without too many essential differences.

Meanwhile, the selections of indicators system in foreign countries are also different. Compared with the domestic indicator system, it has higher universality, and is more international. It can be targeted for the evaluation selection of important cities in the world, and therefore it attracts more attention and is more influential than the domestic indicator system.

Reference

Giffinger R, Fertner C, Kramar H et al (2007) Smart cities – ranking of European medium-sized cities. http://www.smart-cities.eu/download/smart_cities_final_report.pdf. Accessed 10 June 2015

Chapter 7
Principles and Methods of the Intelligent City Evaluation System

Abstract The intelligent city is not a terminal stage, but an on-going and up-grading process that promote the city to be more intensive, smart, green, and low-carbon. Thus, it is important to build a scientific and reasonable intelligent city evaluation indicator system to improve the city's intelligent status. Based on a systematic review of all presented intelligent city evaluation systems (ICESs) from around the world, five primary dimensions of the intelligent city evaluation system are selected, which are Environment and Urbanism, Governance and Public Service, Economy and Industries, Informatization, Innovation Human Resource. From this basis, 5 sets of indicators which are most representative are formulated. The final intelligent city evaluation indicators system is divided into two levels with 5 primary indicators and 20 secondary indicators. In total, 275 experts from 14 high-end R&D institutions participated in the process of creating intelligent city evaluation indicators system.

Keywords Intelligent city · Evaluation indicator system · Methodology Indicator selection

1 Goal of Constructing Intelligent City Evaluation System

An ideal intelligent city is a life entity, and with the support of the information technology, it can sense the outer world, make a judgement, take reactions and even learn new contents. The intelligent city is not a terminal stage, but an on-going and up-grading process that promote the city to be more intensive, smart, green, and low-carbon.

The meaning of the word "intelligent" in the intelligent city is opposite to that of the word "mechanical", which is relevant to the rough development model of the city in the past. With constant breakthroughs and continuous upgrading of big data, cloud platform and Internet of Things, the advance of the technology has led the development of the city from the real space to the digital virtual space. Does it mean that cities that take advantage of these new technologies can be called "intelligent

© Springer Nature Singapore Pte Ltd. and Zhejiang University Press 2018   101
Z. Wu, *Intelligent City Evaluation System*, Strategic Research
on Construction and Promotion of China's Intelligent Cities,
https://doi.org/10.1007/978-981-10-5939-1_7

cities"? The fundamental difference between the development model of traditional cities and that of intelligent cities is that the latter is based on the respect to the environment. It's a kind of wise development, through the systematized, full life cycle development idea, to minimize the consumption of the resources.

The development of city has entered an era led by the information technology. Countries around the world are all facing with the same problem: the basis of human survival is becoming more and more weak, and the problems exposed in the urban environment is constantly increasing, such as: environmental degradation, unreasonable urban expansion, soared unstable social factors, and contradictions between economic development with other aspects of urban development, etc. In the past, people used to take the increase of GDP as the only means to measure the development of a city. Although it's simple and practical, it only measured the increase of economy in quantity, and neglected the changes in quality, and turned a blind eye to the consumption of resources, environment pollution and social justice etc.

Thus, it is vitally important to build a scientific and reasonable city evaluation indicator system. Furthermore, as the information technology has an overall impact on the construction of the city, the development of the city shall be changed from the previous mode. As a conclusion, it is rather important to build an intelligent city evaluation indicator system.

The construction goal of the intelligent city evaluation indicator system is to improve the city's intelligent status, and promote the sustainable development of smart cities through the evaluation of the aspects that can show the level of the city's intelligent development, including economy, society, ecology, urban construction, etc.

2 Relativity of Intelligent City Evaluation Criteria

The key to evaluate the construction status of intelligent city is how to determine the evaluation criteria, namely, to use what kind of indicator to evaluate the construction level and development changes of the intelligent city. Given the fact that cities are widely distributed all over the world, and the natural conditions, economic states, social nature and other situations are all varying from each other, thus it's difficult to build a universal standard for evaluation. In other words, there is no definite evaluation standard, and any standard has its relativity and is always proposed based on real states. Any standards have its own regional, social and historical limitations and there is no evaluation standard without these limitations.

The determination of evaluation standard is up to determine the goal of the evaluation. If the goal is to evaluate and compare the intelligent city construction in different geographical environments, then we can choose the same indicator data of different regions during the same period as the evaluation standard. If the goal is to evaluate the variation of certain region's construction level, then we can choose the same indicator data in many time quantum as the evaluation standard.

We think that the goal of intelligent city development construction evaluation is to understand the situation of intelligent cities states around the world, and judge the development phase of a certain city in a scientific and rational manner, and its pros and cons in nature, economy, society, soft and hard wares, etc. Therefore, for a certain city, it's very important to have a horizontal comparison with the intelligent cities around the world as well as to have a vertical comparison with itself to see it's intelligence trend.

3 Objectives of Intelligent City Evaluation System

The intelligent city evaluation system has been used to evaluate any city that puts forward the construction plan for intelligent city. In the R&D process of this system, we have made trial evaluation to the prefecture-level cities of pilot smart cities selected by the Ministry of House and Rural-urban Development, to investigate its intelligent development degree. Besides, we have made trial evaluation to the key words that have been put forward all over the world, such as "smart", "intelligent" "informatization", and "digitization", etc.

And we made comparison to the results before and after. And use this as the basis to determine the level of our country's intelligent city construction level in global competition.

This evaluation system mainly focused on their most important dimensions in the development of cities, namely, the development environment, future trend, and construction and operation. The indicators been chosen in these three dimensions can differentiate this evaluation system to those for sustainable development cities, ecological cities and innovation cites.

1. Evaluate city's "development environment"

To evaluate how a city deal with its natural, economic, cultural, social and ecological issues, the relevant series of indicators can finally formulate the most important contents of the cities that have been concerned in the intelligent city evaluation system,

2. Evaluate city's "future trend"

To evaluate how local government, treat the future development of a city, whether its strategic judgement can meet the target of intelligence, and whether its judgements are accurate or not. Those are also important parts that been concerned in the evaluate system.

3. Evaluate city's "construction and operation"

While evaluating the construction and operation of a city, it's also an important evaluation point whether this city's development is intelligent and well-organized, and able to obtain real effect or not.

4 Constitution of Intelligent City Evaluation System

Indicators of the evaluation system can be divided into three levels, each with its own stresses and particularities. Primary indicators (dimension) stress on the top-level design, aimed to build the platform for intelligent cities and towns, and therefore to guide the design of each level; secondary indicators is to show each specific systems in the city's development, and take control of each side in the intelligent city development; the tertiary indicators are of the specific city affairs, to show the quantitative treatment of executable events.

1. The primary indicators cover the integration of four modernizations in the intelligent urbanization

The primary indicators shall be easy to remember, easy to grab, easy to control and easy to show, and be able to reflect the major aspects in the intelligent city construction process briefly. Each primary indicator shall be consisted with 3–5 secondary indicators, and the weight of each secondary indicator shall be obtained by the experts in each project through the Delphi Method.

$$A = r_1A_1 + r_2A_2 + \cdots + r_nA_n$$

Among them, A is the primary indicator, and A_1, A_2, ..., A_n are the corresponding secondary indicators, and r_1, r_2, ..., r_n are the weight value of each corresponding secondary indicator within the range of [0, 1].

2. Secondary indicators cover all points and encourage development with characteristics

Secondary indicators are combined with the objectives of each urban management institutions to show the overall coverage of these indicators and the guidance to each key system. When combining the figures in the secondary indicators to the primary indicator, it shall have a non-dimensionalized process to all the indicators and use threshold and the maximum value to make upward synthesis calculation. Therefore, it can show the emphasis and characteristics of the intelligent development of each city, and also be able to promote the intelligent city to formulate its own characteristic and brand.

Based on the maximum value of tertiary indicators, compare each indicator and its threshold value. If all tertiary indicators have achieved the threshold value, then the tertiary indicator with the maximum value shall be chosen. Synthesis calculation of secondary indicators:

$$A_3 = \max(a_{31}, a_{32}, \ldots, a_{3n}) + \mathrm{IF}((a_{31} - a_{31v}) < 0, a_{31} - a_{31v})$$
$$+ \mathrm{IF}((a_{32} - a_{32v}) < 0, a_{32} - a_{32v}) + \cdots + \mathrm{IF}((a_{3n} - a_{3nv}) < 0, a_{3n} - a_{3nv})$$

In this equation, A_3 is the third secondary indicator under the primary indicator A, a_{31}, a_{32}, ..., a_{3n} are n tertiary indicators under A_3, a_{31v}, a_{32v}, ..., a_{3nv} are n threshold values of those tertiary indicators. The function IF(c, x) means if the condition c is satisfied, then use the value x.

3. Open setting of tertiary indicators to guide urban innovation and development

Tertiary indicator system shall be integrated with the quality of people's livelihood of urban and rural residents. It shall have a non-dimensionalized process to each secondary and tertiary indicator, and formulate data within the unified interval between 0 and 1. The setting of tertiary indicators shall have certain openness. In the different phase of the process of intelligent city construction, the adjustment and optimization shall lead the innovative development of intelligent city.

The composition the tertiary indicators to the secondary indicators: set a standard threshold as standard line, if all the indicators are above the standard line, then use the maximum tertiary index, or use the maximum index to minus the difference value.

5   Research and Development Methods of Intelligent City Evaluation Systems

5.1   Determine the Core Guidelines

The core of intelligent city evaluation system is based on the philosophy of urban evolution and the values behind the intelligent urban development trend. Perception, judgement, response and study are four necessary stages of intelligent city development, which composite the rule to guide the construction of the whole system and promote the sustaining evolution of intelligent city in a circular manner (see Fig. 1).

1. Perception: Based on comprehensive perception

The intelligent city with strong perception infrastructure system can master the needs and changes of each city bodies thanks to the support of sufficient data. First of all, an intelligent city shall be sensible in every way, thus it can obtain the required city information and data. With the help of RFID, infrared sensors, global positioning system, laser scanner, etc., the intelligent city can collect the needed information and use sensor network, communication network, mobile Internet to communicate information, so as to sense the city's information normatively, accurately and broadly.

Perception is the basis for the communication between the objects within the city.

2. Judgement: Be able to have an accurate judgement

Any possible state and consequences can be predicted on the basis of experiences and model inference in the intelligent city. City is a complex giant system.

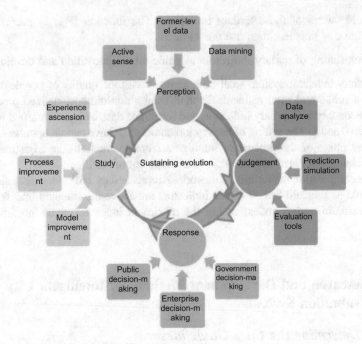

Fig. 1 Basic concept model of intelligent city

The information sensed by the Internet of Things tends to be huge amount of data. The nature, characteristics of energy levels that have been reflected from that huge amount of data are different from each other. The intelligent city can sense those huge data, and through automatically identification of those data with the help of certain tools, skills and methods, to make analysis, calculation and judgement of those information and data to select useful information. The judgement of intelligent city to that huge amount of information is actually the selection process of obtained information.

3. Response: stress on appropriate reflection

Intelligent cities are able to mobilize resources to realize the minimum consumption of energy, resources, time and social psychology based on analysis on the city's development. Intelligent city can not only sense and select the information, but also make intelligent analysis and solution to the obtained information. The intelligent city takes advantage of the selected information, and use the existing knowledge and experience to decide the future action and plan and send control instruction to the relevant actuating equipment.

4. Study: necessity of continuous learning

The intelligence of intelligent city is that it can be improved through the constant "sense-judge-respond" process, and update its judgement model and procedure to realize the continuous advancement of intelligent city. Intelligent city is just like human being which is a life entity. It can make improvement and adjustment to the information procedure, experience based on the overall perception, intelligent judgement and response, and feedback the developed information to the city's system. In an intelligent city, the perception, delivery, analysis and processing of the information is not an end of the process, but the study and development of the information and the process making information more conformed to the social expectation.

5.2 Set Up Primary Indicators

Based on the determined overall construction target of intelligent city, and referring to multiple evaluation systems by relevant institutions and cities domestic and abroad, and the research result of R&D institutions and academicians home and abroad, we summarized all presented indicators (see Table 1). It can be seen from the table that each set of evaluation indicator system has different focus on the construction and operation of the city, and we can conclude the distribution law of each indicator, and get five primary indicators (dimension): Environment and Urbanism, Governance and Public Service, Economy and Industries, Informatization, Innovation Human Resource.

5.3 Set Up Secondary Indicators

For the overall evaluation of intelligent city, based on the 5 chosen framework, namely, Environment and Urbanism, Governance and Public Service, Economy and Industries, Informatization, Innovation Human Resource, we chose typical items from each of those categories to formulate 5 set of indicators which are most representative.

275 experts form 14 R&D institutions[1] participated in the primary indicator system of "urban sustainable development intelligent monitoring", and obtained 220 primary indicators from the summary and induction of indicators on intelligent cities home and abroad (Wu et al. 2011). Since the primary indicators are over-lapped and related, after analyzing those indicators, experts chose the most obtainable and internationalized indicators when two or more indicators are similar

---

[1]Including 14 R&D institutions: Tongji university, Beijing university, Zhejiang university, Shenyang architecture university, Capital normal university, the Ministry of land and resources information center, Ministry of urban and rural planning management center, the State council development research center, National center for remote sensing, Geographic science and resources institute of the Chinese academy of sciences, Chinese academy of sciences institute of remote sensing application, Shanghai urban development information research center, China academy of urban construction, Qingdao prospecting institute of surveying and mapping, etc.

Table 1 Primary indexes distribution

Primary indicator (dimension)	Key words	Frequency
Environment and Urbanism	Natural environment	6
	Smart planning and design	5
	Intelligent building	3
	Transportation innovation and safety	3
	Sustainable recourses utilization	2
Governance and Public Service	Public society service	15
	Government affairs decision-making and management	14
	Medical security	6
	Soft environment	6
	Cultural education	3
	Security	3
	Transparency and public participation	3
	Urban comprehensive functions	2
	Transportation governance	1
	Energy management	1
Intelligent economy and industry	Development of industries	12
	Economic level	5
	Innovation and R&D	5
	Labor force level	2
Informatization	Network infrastructure	9
	Information public platform	4
	Utilization of ICT	3
	Enterprises informatization	2
	Citizen informatization	2
Innovation human resources	Quality of human resources	6
	Education and study	4
	Social cohesion	3
	internationalization	2
	Diverse and open	2

or over related, and abandon the rest. After the cross analysis of those 220 primary indicators, finally 36 preliminary secondary indicators were selected (see Table 2).

5.4 Amendments on the Basis of Experts Advises

After the formation of preliminary secondary and tertiary indicators, we need to merge different indexes together. In the broad Delphi expert consultation, our

Table 2 Preparatory 36 secondary indexes

Primary indicator	Secondary indicator
Intelligent environment and construction	Per capita housing area of urban residents
	Construction land area
	Residential land
	Industrial land
	Green land
	The synthesis pollution index of water environment
	Per capita water resources
	Per capita agricultural acreage
	Per capita construction area
	Natural ecological land coverage rate
	Water supply penetration rate
	Sewage treating rate
	Road hardening rate
	Clean energy penetration rate
	Waste collection rate
Intelligent management and service	Migrant workers endowment insurance coverage rate
	Migrant workers employment injury insurance coverage rate
	Labor dispute cases settlement rate
	Petition cases settlement rate
Intelligent economy and industry	GDP
	Urban labor productivity
	City output density
	The proportion of second industry in GDP
	The proportion of third industry in GDP
	Price of land
Smart hardware facilities	Data network penetration rate
Residents' intelligent potential	Net migration rate
	Total migration rate
	Population structure influence index
	Society influence index
	Resources and environment influence index
	Public service influence index
	Labor market coverage ratio
	Urban-rural income gap
	Employees non-agricultural level
	Per capita energy consumption

research group have sent 56 questionnaires to the academicians and experts in CAE within the project team of "Chinese Intelligent City Construction and Promotion Strategy Study" to make amendment to the indicators, and chosen the indexes with experts' score on each index, and then made appropriate addition and deletion to the specific indexes to solve the match problem within the evaluation system (The detailed content of expert consultation questionnaire can be found in the second part of the appendix).

5.5 *Operational Feedbacks and Indicator Upgrading*

To build up an intelligent evaluation system, we must choose a series of influential and intelligent cities to have a trial operation of this system. Through the process and with the result of the trial operation, we have found and eliminated those less comparative indexes with the passage of time and development of technology, and added more indexes that have the time and social significance (see Fig. 2), and formulated 20 more new secondary indicators (see Table 3).

In the R&D process of intelligent city evaluation process, the data obtained has been transferred from the obtainable government official's statistical material to the public data from the companies and data obtained through information technology Internet (The detailed content of intelligent city evaluation indicators can be found in the third part of the appendix).

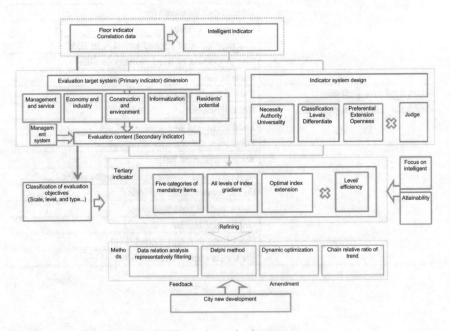

Fig. 2 The process of creating intelligent city evaluation system

Table 3 New 20 secondary indexes

Category	Secondary indicator
Intelligent construction and environment	Density of PM2.5/PM10 monitoring stations in the city
	City grid management level of coverage
	Residents' intelligent transportation tools usage level
	Online publishing level of city future construction plan
Intelligent management and service	Non-classified government document online openness
	Online public participation ratio
	Utilization level of citizen's e-health documents
	Intelligent treatment of emergencies
Intelligent economy and industry	weight of R&D expenses in GDP
	Urban labor productivity
	City output density
	Proportion of city's intelligent industries
Smart hardware facilities	Coverage of free WIFI in public area
	Mobile Internet usage per capita
	City broadband speed
	Intelligent grid level of coverage
Residents' intelligent potential	Proportion of city Internet users
	Proportion of information practitioners
	The proportion of junior college or above educational level population
	Residents' per capita online shopping expenditure amount

6 Interpretation to Intelligent City Evaluation Indicators

The definition of intelligent city in our concept is made up with five parts, namely, Environment and Urbanism, Governance and Public Service, Economy and Industries, Informatization, Innovation Human Resource, and with the frequency statistics, theoretical analysis, Delphi method and reference existing relative evaluation index system of research methods, the preliminary intelligent city evaluation system can be acquired. After we have determined the preliminary intelligent city evaluation system, we can use it to evaluate the present intelligent city construction process. We shall take the economic situation and characteristics of the cities that are involved in the trial evaluation into consideration, and focus on the attainability of the indicators, and then gradually determine the resources of the data, and also make amendments to the indexes. This intelligent city evaluation indicators system is divided into two levels with 5 primary indicators and 20 secondary indicators.

The obtained data for evaluation was partly from the government released statistical material, and partly from the intelligent platform of public Internet, including the statistical yearbook of China and statistical yearbook of urban construction online, and major Internet data information platforms. After the trim and calculation of the data of indicators, and after a non-dimensionalization process to each secondary and tertiary indicator, data within the unified interval between 0 and 1. Then rank and analysis can be made according to final scores (The detailed content of data resources and grading can be found in the fifth and sixth parts of the appendix).

6.1 Indicators for Intelligent Environment and Construction

1. Density of PM2.5/PM10 monitoring stations in the city: The density of air quality PM2.5 or PM10 real-time monitoring points in the city can reflect its perception level to the environmental quality.
2. Coverage of city grid management: The proportion of the regions that have been divided into single grid unit according to certain norm about the whole urban management jurisdiction, can reflect the city's digitalized management level.
3. Citizens utilization rate of intelligent transportation: The degree of citizens using intelligent vehicle and its auxiliary system, such as bus query system, real-time traffic system, etc.
4. The openness degree of the city's future building plan published online: The openness degree of city's building plan released on the government's website can reflect its openness of the intelligent city construction.

6.2 Indicators for Intelligent Management and Services

1. Non-classified government document online openness: The percentage of the non-classified documents in the total amount of documents can reflect the transparency of government's information.
2. Online public participation ratio: The proportion of public participation in the decision of city construction related events. It can reveal the citizens' participation degree of city construction, and the openness, fairness and inclusiveness of decisions.
3. Utilization level of citizen's e-health documents: The percentage of citizens with e-health documents can reflect the digitization degree of citizen's information.
4. Intelligent emergency response: The function level of intelligent emergency system in face of city's major emergencies (such as disaster, accidents, etc.)

6.3 Indicators for Intelligent Industries and Economy

1. Weight of R&D expenses in GDP: The percentage of R&D expenses in GDP reflects the technology strength, innovation ability and core competitiveness.

2. Urban labor productivity: Per capita GDP reveals local economic development.
3. City output density: The average GDP created by each square kilometer fully reflects the wisdom of land utilization.
4. Proportion of city's intelligent industries: The proportion of city's knowledge-intensive and technology-intensive industries to the whole city's industries.

6.4   Indicators for Intelligent Hardware Construction

1. Coverage of free WIFI in public area: The proportion of areas that provide free WIFI in the whole city space, it can reveal the accessible level of information from the hardware aspect.
2. Mobile Internet usage per capita: The usage rate of mobile Internet (mobile phone 3G/4G, etc.) can reflect the mobile Internet construction level.
3. City broadband speed: The city's broadband speed is one of the basic link of the intelligent city construction.
4. Coverage of smart power grid: The coverage of smart power grid can reflect the intelligent level of the city's energy.

6.5   Indicators for Residents' Intelligent Potential

1. Proportion of city Internet users: The percentage of Internet users can reflect the citizen's information attainability and study ability.
2. Proportion of information practitioners: The proportion of information practitioners in all the city workers.
3. Proportion of people with college degree and above: The proportion of people with college degree or above can reflect the city's intelligent degree from the citizens' education level.
4. The amount of citizens' consumption online per capita: The proportion of online consumption per capita to the total consumption per capita can reflect the popularity of Internet and development degree of Internet of Things indirectly.

7   The Process of Creating Intelligent City Evaluation System

In the process of creating Intelligent City Evaluation System, the research group has held many academician expert seminars (see Table 4), and listened to the suggestions from all fields to discuss on the framework and the indexes of the evaluation system. The academician experts in CAE's "Chinese Intelligent City Construction and Promotion Strategy Study" project team have put forward many valuable suggestions to this evaluation indicator system.

Academician Pan Yunhe emphasized in the seminar that the evaluation indexes shall be based on the integration of researches home and abroad, and discussed with experts from each fields. Through this, the indicator system can be capable for the comparison of intelligent cities between domestic and foreign countries, and can demonstrate the overall achievement of each field in the construction process of intelligent cities. Therefore, it can be able to assess the intelligent city in a better way, and construct the intelligent city more well-directed. This is just why this evaluation system take lead of the evaluation systems put forward by any single city or institution.

Academician Xiang Haifan believes that if we are going to make a conclusion of China's city development mode, it shall put emphasis on green and intelligence in the intelligent city construction process.

Academician Fan Lichu holds that, since the quality and function of each city varies, the focuses of intelligent city construction differ from each other, this evaluation system shall be able to demonstrate the differences between different types of cities' intelligent construction.

Table 4 The process of creating intelligent city evaluation evaluation system

Time	Seminar
Mar. 21, 2013	"Intelligent City Standard System Building Study" seminar in Tongji University
Sept. 10, 2013 to Sept. 16, 2013	CAE representative team visited Munich to attend the first SIno-Germany intelligent city development seminar that jointly held by CAE and ACATECT
Sept. 27, 2013	"Intelligent City Evaluation Indicator Systems "promotion seminar held in Qiandao Lake
Nov. 31, 2013	"Intelligent City Evaluation Indicator Systems "promotion seminar held in Tongji University
Ari. 21, 2014	"Intelligent City Evaluation Indicator Systems" conclusion seminar held in CAE
Aug. 29, 2014	Professor Wu Zhiqiang made a preach on intelligent city evaluation indicator system in "2014 China Shanghai intelligent city innovation development summit"
Oct. 16, 2014	Professor Wu Zhiqiang made a preach on intelligent city evaluation indicator system in "Global intelligent city summit forum"

Academician Dai Fudong thinks that people is very important in the intelligent city construction, thus it shall be reflected in the selection of indexes. The target of intelligent city shall be not only the application of intelligent technology, but also the improvement of citizens' intelligent potential.

Academician Jiang Huancheng thinks that this evaluation indicator system shall show the Chinese characteristics. He suggested demonstrating the city's intelligence construction from the necessities of life.

Academician Li Baotong believes that the selection of intelligent city evaluation indicators shall take the differences of each city into consideration. In this way, we can make comparison between cities with different backgrounds. Besides, it's very important to collect the data of each indicator, the data shall not from the city's statistical statement, but obtained through technological means.

Academician Jiang Yi holds that the intelligent city evaluation system will promote the construction of the construction of intelligent cities in China. He also suggested adding the index of city spatial information, because the basic information such as the city's geographical data is the foundation for the construction of intelligent cities.

Academician Wu Manqing believes that the intelligent city evaluation system shall be pre-released in the R&D process, so the broad criticism and discussion from the city's management level as well as citizens can be obtained. In this way, the evaluation indexes can be amended and the influence of the system can also be expanded. For the authenticity of the data, it can refer to some official data, and analyze with the data released officially.

Academician Yu Yixin thinks that each branch research team of "Chinese Intelligent City Construction and Promotion Strategy Study" project team can

Fig. 3 A photo of Professor Wu Zhiqiang introducing intelligent city evaluation indicator system in "Global intelligent city summit forum"

provide some evaluation method and content of this system, such as the evaluation of Smart power grids can contain more than 20 indexes.

After many academician seminars in the period of 2013–2014, the intelligent city evaluation system has been perfected based on multiple times of trial evaluation, and expanded from domestic evaluation to Europe and America. It enabled us to stand on the international perspective to see the pros and cons of city construction in our country, and clearly see the gap between us and the leading international cities. The intelligent city evaluation system has been released through the announcement in International conferences, and had wide repercussions (Fig. 3).

Reference

Wu ZQ, Qiu YY, Gan L et al (2011) Scientific and rational support key points of urbanization in China - Ministry of Science and Technology "11th Five-Year Plan" Science and Technology Support Projects - "The Dynamic Monitoring Key Technologies of Urbanization and Rural Construction" Overview. City Plan J (4):1–9

Chapter 8
Construction Level Ranking of Intelligent Cities

Abstract From 2013 to 2014, 33 cities from the list of wisdom city pilots released by the Ministry of Housing and Urban in China and 33 American and European cities who have a certain basis for the development of intelligent cities were selected to conduct the trial evaluation. This chapter presents a deep interpretation on the evaluation results. In addition, it provides a new approach to describe the growth and development stage of global intelligent cities, which divides the 41 cities at home and abroad into four quadrants based on the level of intelligent cities' growth and development respectively. For the future, the global intelligent cities construction is expected to have more worldwide in-depth cooperation, pay more attention from hardware to social quality and expand from a single city to city agglomeration area.

Keywords Intelligent city · Evaluation indicator system · World
Growth and development stage · Trend

1 Construction Level Ranking of Intelligent Cities in China

From 2013 to 2014, the Intelligent City Evaluation System has passed a number of trial assessments. According to the result, the group researching on this topic adjust the evaluation indicators. During the trial evaluation process, participated cities are totally 193, which are announced by the Ministry of Housing and Urban in 2012 and 2013 to be the two batches of national wisdom city pilots. At first, the group researching on this topic chose 33 prefecture-level cities from the list of wisdom city pilots to conduct the trial evaluation (The criteria for selecting the evaluation object are that the selected cities have a certain basis for the development of intelligent cities).

According to the Constitution of Intelligent City Indicator System in Chap. 7, we get the comprehensive and sub-item evaluation of the 33 trial evaluation cities as shown in Table 1 and Fig. 1. The top five cities of the comprehensive evaluation

© Springer Nature Singapore Pte Ltd. and Zhejiang University Press 2018   117
Z. Wu, *Intelligent City Evaluation System*, Strategic Research
on Construction and Promotion of China's Intelligent Cities,
https://doi.org/10.1007/978-981-10-5939-1_8

are Jinhua, Ningbo, Zhuhai, Wenzhou and Wuhan. Each city has its own features, and the detailed data sources and scores are in the fifth and sixth parts of appendix.

The above results show that leading cities' comprehensive scores are all above 50, such as Jinhua, Ningbo, Zhuhai, Wenzhou, Wuhan, Nanjing, Wuxi, Shanghai Pudong, Taizhou, Changzhou, Weihai, Zhenjiang, Dongying and Langfang, and Jinhua is the first with the score of 62.92 points while Wuhai and Liaoyuan are the cities with the lowest score of only around 20 points. We can see that most of the top cities are all in China Eastern coastal areas while many Midwestern cities are low-ranked. Thus, although these cities are all announced to be the National Wisdom City Pilots by the Ministry of Housing and Urban, there are great gaps among their actual developments of the Intelligent Construction.

In addition, the highest score (62.92 points) in the comprehensive evaluation is nearly 3 times as many as the lowest (21.55 points). Such a disparity shows that 5 major indicators in Constitution of Intelligent City Indicator System are consistent with the public's feeling of the Construction of Intelligent City and these indicators are also sensitive in each determinations of the Construction of Intelligent City.

The above 33 cities' evaluation of Intelligent Environment and Construction, Intelligent Management and Services, Intelligent Economy and Industries, Intelligent Hardware Facilities and Residents' Intelligent literacy is shown in Tables 2, 3, 4 5 and 6 and Figs. 2, 3, 4, 5 and 6.

In the matter of Intelligent Environment and Construction (see Table 2 and Fig. 2), Dongying, Weihai and Jinhua get the highest score above 85 points while Nanping, Liaoyuan and Wuhai are with the lowest score below 40 points. Although the government put huge investment in the Construction of Intelligent Environment in the eastern developed regions, because of their environment being in a state of high pollution, there are no outstanding achievements in Intelligent Environment and Construction, while some of the coastal areas of medium-sized cities (such as Dongying and Weihai) are with better environmental quality as well as in the leading place.

In the matter of Intelligent Management and Services (see Table 3 and Fig. 3), Shanghai Pudong, Nanjing and Ningbo get the highest score above 70 points while Handan, Langfang are with the lowest score below 25 points. These show that there is still great room for improvement in intelligent management of the government and the intelligent services of the social life for many cities, and by the Construction of Intelligent cities, the modernization of urban social governance can be accelerated.

In the matter of Intelligent Economy and Industries (see Table 4 and Fig. 4), except for Lhasa, the highest score is nearly more than 4 times as many as the lowest. It indicates that after the past 30 years of economic reform, the cities whether in the developed coastal areas or inland areas, have already got relatively sufficient understanding of the intelligent technology's contribution to promoting the local economic development. At the same time, the intelligent information technology that is used to promote the development of the local economy will spread at a faster rate in other cities.

Table 1 Comprehensive and sub-item evaluation of intelligent construction of the 33 cities in China

City	Total		Intelligent environment and construction		Intelligent management and services		Intelligent economy and industries		Intelligent hardware facilities		Residents' intelligent literacy	
	Ranking	Score	Ranking	Score	Ranking	Score	Ranking	Score	Ranking	Score	Ranking	Score
Jinhua	1	62.92	3	86.08	28	36.50	3	56.17	1	77.91	1	57.93
Ningbo	2	57.09	6	82.65	3	70.50	23	32.29	6	58.79	8	41.20
Zhuhai	3	56.13	14	71.31	7	61.47	27	27.70	2	69.84	4	50.33
Wenzhou	4	55.80	8	81.14	26	39.98	20	33.60	3	68.58	2	55.67
Wuhan	5	55.44	9	79.84	4	69.29	24	31.18	9	54.55	7	42.33
Nanjing	6	54.78	18	66.35	2	74.94	16	35.16	5	59.02	9	38.45
Wuxi	7	54.67	7	81.60	11	60.22	2	57.65	14	46.50	22	27.41
Shanghai Pudong	8	54.48	11	75.45	1	75.65	11	40.45	13	50.61	18	30.24
Taizhou	9	53.98	19	66.19	13	58.66	1	66.51	18	42.83	11	35.72
Changzhou	10	53.25	16	70.37	21	46.38	4	52.27	4	64.39	15	32.84
Weihai	11	53.22	2	87.11	14	55.57	9	43.67	16	44.75	13	34.99
Zhenjiang	12	53.09	5	84.55	22	46.16	5	51.68	12	51.60	16	31.46
Dongying	13	51.97	1	87.50	10	60.28	7	48.44	27	28.44	12	35.19
Langfang	14	51.85	4	84.98	32	23.16	13	40.11	8	56.16	3	54.81
Dezhou	15	48.41	13	73.02	9	60.55	8	44.31	19	41.25	24	22.90
Xianyang	16	47.65	10	78.78	16	50.41	18	34.43	23	36.95	10	37.66
Yaan	17	45.77	23	50.00	24	42.49	15	35.16	10	53.11	6	48.09
Nanpin	18	45.47	33	25.00	18	48.96	6	51.64	11	52.45	5	49.29
Zhuzhou	19	43.61	21	63.57	12	59.59	25	29.91	22	37.01	21	27.95
Tongling	20	42.75	24	50.00	25	40.34	14	37.28	7	56.22	19	29.91
Wuhu	21	42.69	22	57.49	19	46.55	12	40.26	20	40.73	20	28.42

(continued)

Table 1 (continued)

City	Total		Intelligent environment and construction		Intelligent management and services		Intelligent economy and industries		Intelligent hardware facilities		Residents' intelligent literacy	
	Ranking	Score	Ranking	Score	Ranking	Score	Ranking	Score	Ranking	Score	Ranking	Score
Lasa	22	40.82	17	67.33	17	49.49	33	7.90	17	44.58	14	34.78
Changzhi	23	40.75	20	63.78	6	63.67	28	25.13	26	30.92	26	20.25
Bengbu	24	40.43	25	50.00	8	61.23	17	34.72	21	38.58	27	17.63
Huainan	25	40.36	12	75.00	23	44.05	22	32.41	24	35.65	29	14.67
Pingxiang	26	38.67	26	50.00	29	34.62	10	43.18	15	44.78	25	20.76
Hebi	27	37.57	27	50.00	5	67.73	26	28.72	28	27.88	30	13.55
Qinhuangdao	28	36.60	30	49.33	15	55.02	30	20.53	29	27.47	17	30.67
Handan	29	35.19	15	70.97	33	20.79	19	34.25	30	23.88	23	26.06
Liupanshui	30	33.03	28	50.00	20	46.48	29	22.92	25	34.81	33	10.92
Luohe	31	30.22	29	50.00	31	31.14	21	33.35	31	19.83	28	16.77
Wuhai	32	22.48	31	37.50	27	38.93	32	16.32	33	6.91	31	12.72
Liaoyuan	33	21.55	32	37.50	30	32.50	31	17.56	32	9.23	32	10.95

The calculation principle in scoring is shown in the Affix III, IV and V

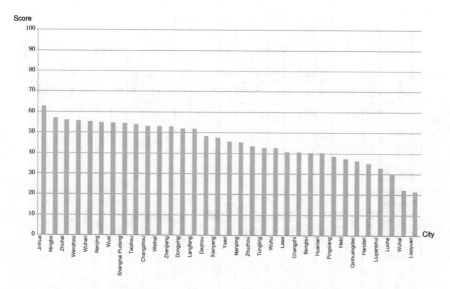

Fig. 1 Comprehensive score of intelligent construction of the 33 cities in China

In the matter of Intelligent Hardware Facilities (see Table 5 and Fig. 5), the higher scores were highly correlated with the hardware facilities in the cities, and cities with the highest scores above 60 points are Jinhua, Zhuhai, Wenzhou and Changzhou, whose scores are more than ten times as many as the lowest. Therefore, it's easy to see that the overall ranking of wisdom cities is highly affected by the intelligent hardware facilities.

In the matter of Residents' Intelligent literacy (see Table 6 and Fig. 6), the cities with the highest scores above 50 points are Jinhua, Wenzhou, Langfang and Zhuhai, while the cities with the lowest scores below 10 points are Liupanshui and Liaoyuan. The score gap in this part has been enough to raise a great concern of the local government to vigorously enhance the urban Residents' Intelligent Literacy.

Generally speaking, Intelligent City Evaluation Indicator System has gone over the pure social, economic and environmental indicator system which was separated from intelligent technology, now closely connected the intelligent elements with the sustainable development of the city's society, economy and environment indicators. If this evaluation system was officially launched, a comprehensive evaluation can be conducted to nearly 200 pilot intelligent or wisdom cities.

The group researching on this topic suggests working together with the European scientific research as soon as possible in the next phase to jointly promote the trial evaluation of the European wisdom city. At the same time, we shall conduct a domestic trial evaluation about the construction of the wisdom city and wisdom development zone, and launch the corresponding evaluation system of intelligent city and intelligent industry park as soon as possible.

Table 2 The evaluation of intelligent environment and construction of the 33 cities in China

Ranking	City	The density of city PM201/PM100 monitoring sites	The level of the coverage of city network management	The level of citizens' usage of intelligent traffic tools	The level of the publish online of the city's future development plan	Comprehensive score
1	Dongying	100.00	100.00	100.00	50.00	87.50
2	Weihai	48.44	100.00	100.00	100.00	87.11
3	Jinhua	44.34	100.00	100.00	100.00	86.08
4	Langfang	89.92	100.00	100.00	50.00	84.98
5	Zhenjiang	38.21	100.00	100.00	100.00	84.55
6	Ningbo	30.60	100.00	100.00	100.00	82.65
7	Wuxi	26.39	100.00	100.00	100.00	81.60
8	Wenzhou	74.57	50.00	100.00	100.00	81.14
9	Wuhan	19.37	100.00	100.00	100.00	79.84
10	Xianyang	65.11	50.00	100.00	100.00	78.78
11	Shanghai Pudong	1.80	100.00	100.00	100.00	75.45
12	Huainan	0.00	100.00	100.00	100.00	75.00
13	Dezhou	92.10	50.00	100.00	50.00	73.02
14	Zhuhai	35.26	50.00	100.00	100.00	71.31
15	Handan	83.88	50.00	100.00	50.00	70.97
16	Changzhou	31.47	50.00	100.00	100.00	70.37
17	Lasa	69.33	100.00	50.00	50.00	67.33
18	Najing	15.38	50.00	100.00	100.00	66.35
19	Taizhou	64.77	50.00	50.00	100.00	66.19
20	Changzhi	55.13	100.00	50.00	50.00	63.78
21	Zhuzhou	54.30	50.00	100.00	50.00	63.57

(continued)

Table 2 (continued)

Ranking	City	The density of city PM201/PM100 monitoring sites	The level of the coverage of city network management	The level of citizens' usage of intelligent traffic tools	The level of the publish online of the city's future development plan	Comprehensive score
22	Wuhu	29.96	100.00	50.00	50.00	57.49
23	Yaan	0.00	50.00	50.00	100.00	50.00
24	Tongling	0.00	50.00	50.00	100.00	50.00
25	Bengbu	0.00	50.00	100.00	50.00	50.00
26	Pingxiang	0.00	100.00	0.00	100.00	50.00
27	Hebi	0.00	50.00	50.00	100.00	50.00
28	Liupanshui	0.00	50.00	100.00	50.00	50.00
29	Luohe	0.00	50.00	100.00	50.00	50.00
30	Qinhuangdao	47.32	50.00	50.00	50.00	49.33
31	Wuhai	0.00	50.00	50.00	50.00	37.50
32	Liaoyuan	0.00	50.00	0.00	100.00	37.50
33	Nanping	0.00	50.00	0.00	50.00	25.00

Table 3 The evaluation of intelligent management and services of the 33 cities in China

Ranking	City	The online publicity of the government non secret-related official documents	The participation ratio of the public online	The level of the citizens' usage of the electronic health record	The level of the emergency intelligent reaction	Comprehensive score
1	Shanghai Pudong	100.00	2.59	100.00	100.00	75.65
2	Nanjing	46.99	52.78	100.00	100.00	74.94
3	Ningbo	50.35	31.64	100.00	100.00	70.50
4	Wuhan	51.39	25.77	100.00	100.00	69.29
5	Hebi	40.96	29.94	100.00	100.00	67.73
6	Changzhi	95.45	9.24	50.00	100.00	63.67
7	Zhuhai	91.07	4.83	50.00	100.00	61.47
8	Bengbu	42.23	2.70	100.00	100.00	61.23
9	Dezhou	42.20	100.00	50.00	50.00	60.55
10	Dongying	30.76	10.35	100.00	100.00	60.28
11	Wuxi	48.42	42.45	50.00	100.00	60.22
12	Zhuzhou	29.34	9.02	100.00	100.00	59.59
13	Taizhou	39.80	94.83	50.00	50.00	58.66
14	Weihai	35.22	37.08	50.00	100.00	55.57
15	Qinhuangdao	40.07	30.00	100.00	50.00	55.02
16	Xianyang	50.26	1.36	100.00	50.00	50.41
17	Lasa	8.39	39.59	100.00	50.00	49.49
18	Nanping	41.67	4.18	100.00	50.00	48.96
19	Wuhu	84.62	1.57	100.00	0.00	46.55
20	Liupanshui	35.94	0.00	100.00	50.00	46.48

(continued)

Table 3 (continued)

Ranking	City	The online publicity of the government non secret-related official documents	The participation ratio of the public online	The level of the citizens' usage of the electronic health record	The level of the emergency intelligent reaction	Comprehensive score
21	Changzhou	35.40	0.13	100.00	50.00	46.38
22	Zhenjiang	47.54	37.08	100.00	0.00	46.16
23	Huainan	70.80	5.41	100.00	0.00	44.05
24	Yaan	8.39	11.57	100.00	50.00	42.49
25	Tongling	48.46	12.88	50.00	50.00	40.34
26	Wenzhou	44.52	15.41	50.00	50.00	39.98
27	Wuhai	5.59	0.12	100.00	50.00	38.93
28	Jinhua	39.29	6.69	100.00	0.00	36.50
29	Pingxiang	52.12	36.35	0.00	50.00	34.62
30	Liaoyuan	22.37	7.63	50.00	50.00	32.50
31	Luohe	22.37	2.19	100.00	0.00	31.14
32	Langfang	31.32	11.33	50.00	0.00	23.16
33	Handan	27.97	5.18	50.00	0.00	20.79

Table 4 The evaluation of intelligent economy and industries of the 33 cities in China

Ranking	City	The proportion of R&D expenditure in GDP	Urban labor productivity	City output density	Proportion of urban intelligent industry	Comprehensive score
1	Taizhou	3.35	71.79	100.00	90.89	66.51
2	Wuxi	6.86	65.89	66.95	90.89	57.65
3	Jinhua	2.42	100.00	95.49	26.77	56.17
4	Changzhou	5.63	54.78	57.78	90.89	52.27
5	Zhenjiang	3.29	54.93	57.63	90.89	51.68
6	Nanping	1.58	83.22	86.73	35.02	51.64
7	Dongying	1.61	84.93	72.80	34.41	48.44
8	Dezhou	0.20	62.12	80.49	34.41	44.31
9	Weihai	3.54	88.31	48.43	34.41	43.67
10	Pingxiang	1.70	30.64	40.37	100.00	43.18
11	Shanghai Pudong	100.00	23.48	15.23	23.10	40.45
12	Wuhu	22.20	22.05	24.95	91.84	40.26
13	Langfang	1.60	64.02	72.11	22.72	40.11
14	Tongling	5.35	26.56	25.36	91.84	37.28
15	Yaan	0.52	33.62	51.22	55.28	35.16
16	Nanjing	10.46	17.00	22.30	90.89	35.16
17	Bengbu	11.42	17.04	18.59	91.84	34.72
18	Xianyang	0.91	29.49	55.17	52.16	34.43
19	Handan	1.06	34.28	66.12	35.53	34.25
20	Wenzhou	3.74	52.15	51.75	26.77	33.60
21	Luohe	1.25	27.35	38.21	66.61	33.35
22	Huainan	4.21	12.69	20.90	91.84	32.41
23	Ningbo	8.64	42.69	51.08	26.77	32.29
24	Wuhan	5.95	22.03	36.61	60.14	31.18
25	Zhuzhou	5.28	30.30	35.56	48.48	29.91
26	Hebi	1.68	22.43	24.16	66.61	28.72
27	Zhuhai	11.42	35.62	32.68	31.09	27.70
28	Changzhi	1.97	30.49	52.17	15.88	25.13
29	Liupanshui	1.52	38.44	43.71	8.01	22.92
30	Qinhuangdao	1.82	23.63	33.94	22.72	20.53
31	Liaoyuan	1.73	18.32	29.74	20.45	17.56
32	Wuhai	5.26	19.59	20.78	19.66	16.32
33	Lasa	0.00	18.44	9.56	3.60	7.90

Table 5 The evaluation of intelligent hardware construction of the 33 cities in China

Ranking	City	The coverage density of the free network in the public space	The per capita usage of the mobile network	Urban broadband speed	The level of the coverage of the smart grid	Comprehensive score
1	Jinhua	100.00	100.00	61.66	50.00	77.91
2	Zhuhai	23.71	93.10	62.55	100.00	69.84
3	Wenhou	27.28	74.69	72.34	100.00	68.58
4	Changzhou	52.40	38.55	66.60	100.00	64.39
5	Nanjing	21.84	31.56	82.67	100.00	59.02
6	Ningbo	41.19	36.15	57.83	100.00	58.79
7	Tongling	19.50	29.79	75.60	100.00	56.22
8	Langfang	4.72	51.58	68.35	100.00	56.16
9	Wuhan	20.09	43.93	54.18	100.00	54.55
10	Yaan	7.41	55.03	100.00	50.00	53.11
11	Nanping	31.36	74.93	53.53	50.00	52.45
12	Zhenjiang	31.14	29.41	45.86	100.00	51.60
13	Shanghai Pudong	5.14	20.18	77.10	100.00	50.61
14	Wuxi	26.91	35.96	73.13	50.00	46.50
15	Pingxiang	27.16	30.86	71.11	50.00	44.78
16	Weihai	3.50	47.56	77.93	50.00	44.75
17	Lasa	29.59	26.99	21.76	100.00	44.58
18	Taizhou	39.85	18.78	62.68	50.00	42.83
19	Dezhou	1.49	19.13	44.39	100.00	41.25
20	Wuhu	26.90	32.07	53.95	50.00	40.73
21	Bengbu	12.79	14.88	76.66	50.00	38.58
22	Zhuzhou	11.16	24.91	61.97	50.00	37.01
23	Xianyang	15.80	28.12	53.87	50.00	36.95
24	Huainan	48.18	10.95	83.46	0.00	35.65
25	Liupanshui	5.82	40.21	43.23	50.00	34.81
26	Changzhi	2.62	34.70	36.35	50.00	30.92
27	Dongying	0.51	18.85	44.39	50.00	28.44
28	Hebi	1.60	2.33	57.61	50.00	27.88
29	Qinhuangdao	3.72	13.31	42.87	50.00	27.47
30	Handan	0.32	15.61	29.59	50.00	23.88
31	Luohe	0.73	21.00	57.61	0.00	19.83
32	Liaoyuan	1.34	1.95	33.65	0.00	9.23
33	Wuhai	1.48	4.40	21.76	0.00	6.91

Table 6 The evaluation of the residents' intelligent literacy of the 33 cities in China

Ranking	City	The proportion of the internet users	The proportion of information practitioners	The proportion of the population with college level or above	The per capita net purchase expenditure	Comprehensive score
1	Jinhua	46.18	66.50	69.04	50.00	57.93
2	Wenzhou	100.00	22.31	25.38	75.00	55.67
3	Langfang	25.19	44.18	99.86	50.00	54.81
4	Zhuhai	11.27	57.37	57.70	75.00	50.33
5	Nanping	27.29	100.00	44.87	25.00	49.29
6	Yaan	10.30	57.06	100.00	25.00	48.09
7	Wuhan	8.46	21.03	64.82	75.00	42.33
8	Ningbo	18.53	20.72	25.54	100.00	41.20
9	Nanjing	7.28	11.95	59.58	75.00	38.45
10	Xianyang	7.41	21.55	46.70	75.00	37.66
11	Taizhou	17.90	44.87	30.13	50.00	35.72
12	Dongying	12.70	42.48	35.58	50.00	35.19
13	Weihai	18.59	18.23	53.14	50.00	34.99
14	Lasa	0.00	90.05	24.07	25.00	34.78
15	Changzhou	19.27	17.82	44.28	50.00	32.84
16	Zhenjiang	13.29	15.91	46.65	50.00	31.46
17	Qinhuangdao	12.87	16.80	42.99	50.00	30.67
18	Shanghai Pudong	22.13	14.53	9.28	75.00	30.24
19	Tongling	6.97	10.74	26.92	75.00	29.91
20	Wuhu	6.40	8.68	48.61	50.00	28.42
21	Zhuzhou	5.62	22.46	33.71	50.00	27.95

(continued)

Table 6 (continued)

Ranking	City	The proportion of the internet users	The proportion of information practitioners	The proportion of the population with college level or above	The per capita net purchase expenditure	Comprehensive score
22	Wuxi	17.09	17.83	24.73	50.00	27.41
23	Handan	19.00	18.30	16.93	50.00	26.06
24	Dezhou	11.70	26.74	28.17	25.00	22.90
25	Pingxiang	11.24	36.17	10.61	25.00	20.76
26	Changzhi	10.13	26.60	19.29	25.00	20.25
27	Bengbu	5.78	11.40	28.35	25.00	17.63
28	Luohe	8.60	15.15	18.34	25.00	16.77
29	Huainan	3.33	6.42	23.94	25.00	14.67
30	Hebi	8.88	11.35	8.96	25.00	13.55
31	Wuhai	4.48	17.84	3.56	25.00	12.72
32	Liaoyuan	3.62	8.44	6.75	25.00	10.95
33	Liupanshui	7.45	25.38	10.85	0.00	10.92

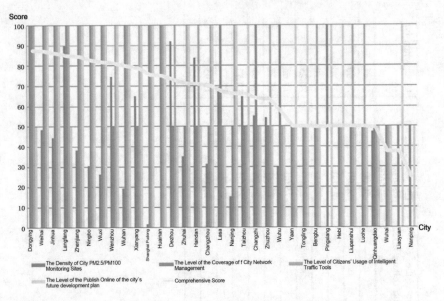

Fig. 2 The score of intelligent environment and construction of the 33 cities in China

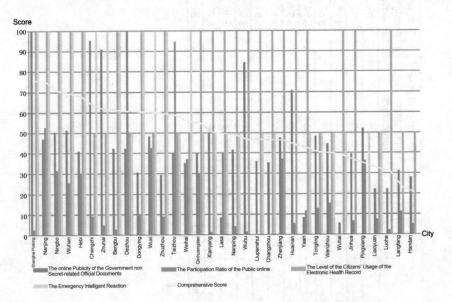

Fig. 3 The score of intelligent management and services of the 33 cities in China

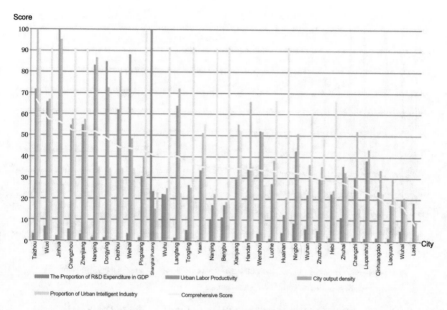

Fig. 4 The score of intelligent economy and industries of the 33 cities in China

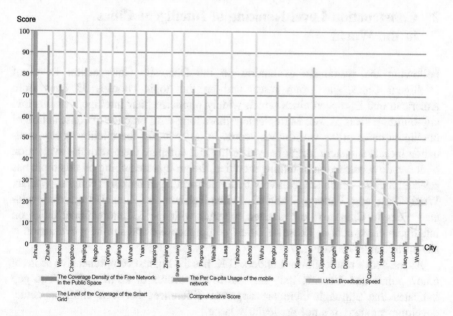

Fig. 5 The score of intelligent hardware construction of the 33 cities in China

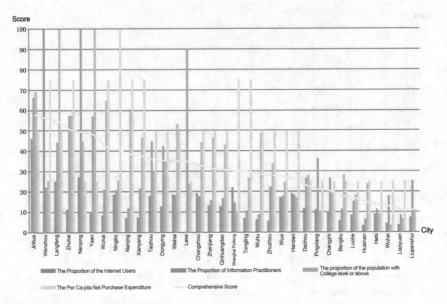

Fig. 6 The score of the residents' intelligent literacy of the 33 cities in China

2 Construction Level Ranking of Intelligent Cities in the World

Following the systematic evaluation on the Domestic Construction Level of Intelligent Cities, the group researched on this topic chooses 33 ones from American and European cities which widely publicize their intelligent or wisdom city-building ideas as well as implement more long-term practice of construction of intelligent cities in the world-wide scope. In order to test the intelligent city evaluation indicator system for global versatility and compare them with the evaluation of domestic construction of intelligent cities, the group researching on this topic has selected 8 cities which are relatively better performing in ranking (Ningbo, Wuhan, Wuxi, Zhenjiang, Pudong Shanghai, Jinhua Taizhou and Zhuhai) from the evaluated cities in China to be evaluated together with the 33 international cities on intelligent construction. These 41 cities' comprehensive and sub-item evaluation is shown in Table 7 and Fig. 7.

As the above result shows, London, U.K. has the highest comprehensive score of 65.67 points and Verona, Italy has the lowest scores of 25.81 points. The gap indicates that although there are apparent differences between cities in actual development state, it's not particularly large.

This also shows that 41 cities selected are exactly the typical ones in the practice of intelligent cities worldwide.

Cities whose comprehensive scores are above 60 points include London, U.K. (65.67 points), Amsterdam, Holland (65.51 points), Helsinki, Finland (64.01

Table 7 The comprehensive and sub-item evaluation of the 41 cities at home and abroad

City	Total		Intelligent environment and construction		Intelligent management and services		Intelligent economy and industries		Intelligent hardware facilities		Residents' intelligent literacy	
	Ranking	Score	Ranking	Score	Ranking	Score	Ranking	Score	Ranking	Score	Ranking	Score
London	1	65.67	13	77.66	4	72.05	6	53.33	4	63.47	5	61.85
Amsterdam	2	65.51	1	97.84	3	72.86	5	54.14	10	56.56	27	46.13
Helsinki	3	64.01	10	84.84	1	74.98	8	47.55	23	47.53	3	65.15
Boston	4	63.87	6	88.42	2	74.36	3	59.63	31	41.29	14	55.65
Copenhagen	5	62.92	8	85.90	27	50.60	4	59.60	5	62.72	13	55.78
Vienna	6	61.22	4	92.03	15	68.98	22	35.83	7	60.97	25	48.30
Washington DC	7	60.92	24	67.79	19	61.54	1	75.58	24	45.85	23	53.84
Seattle	8	60.02	2	92.43	33	45.91	9	47.07	9	59.53	18	55.16
Chicago	9	59.04	18	75.70	13	70.09	17	42.38	17	51.82	17	55.19
San Jose	10	58.77	14	76.86	11	70.16	24	34.57	14	53.58	8	58.67
Portland	11	57.92	23	68.55	10	70.22	20	38.82	12	54.80	10	57.21
San Diego	12	57.09	17	76.00	5	71.60	28	33.11	21	49.69	19	55.06
Dubuque	13	56.62	26	62.50	8	70.95	34	27.43	18	50.19	1	72.04
Manchester	14	56.21	12	82.27	28	48.04	7	52.08	36	35.09	4	63.59
New York	15	55.51	22	72.10	32	45.95	14	44.12	8	60.83	22	54.56
Barcelona	16	55.22	20	75.00	26	53.71	10	46.70	2	68.73	28	31.98
Detroit	17	52.51	34	44.52	12	70.15	18	41.42	19	49.97	11	56.49
Minneapolis and Sao Paulo	18	52.16	27	62.50	20	59.85	15	43.67	35	39.16	15	55.61
Philadelphia	19	52.12	15	76.21	31	46.31	16	43.12	33	40.34	21	54.63
Ningbo	20	51.86	7	86.81	6	71.52	37	22.54	3	65.00	37	13.44
Icy-les-Moulineaux in Paris	21	51.39	38	25.00	23	58.19	30	30.95	1	71.27	2	71.55

(continued)

Table 7 (continued)

City	Total		Intelligent environment and construction		Intelligent management and services		Intelligent economy and industries		Intelligent hardware facilities		Residents' intelligent literacy	
	Ranking	Score	Ranking	Score	Ranking	Score	Ranking	Score	Ranking	Score	Ranking	Score
Francisco	22	50.96	25	66.71	30	47.12	13	45.27	32	40.89	20	54.83
Lisbon	23	49.51	28	62.50	16	67.05	29	31.90	6	62.43	34	23.69
Cleveland	24	48.46	35	37.50	21	59.08	19	40.52	20	49.84	16	55.37
Birmingham	25	47.48	37	28.14	7	71.20	12	45.73	40	32.51	6	59.79
Aarhus	26	47.40	29	62.50	38	25.07	11	45.79	26	45.25	9	58.40
Liverpool	27	46.65	40	20.02	22	58.41	2	61.21	37	34.10	7	59.53
Wuhan	28	46.23	11	82.48	14	69.19	35	27.06	30	41.86	41	10.58
Wuxi	29	45.85	9	85.18	17	62.89	23	35.31	38	33.31	38	12.58
Turin	30	45.33	31	50.00	18	61.84	27	33.97	13	53.60	32	27.26
Zhenjiang	31	45.21	5	89.75	29	47.34	25	34.51	29	42.51	39	11.95
Shanghai Pudong	32	45.06	19	75.70	9	70.56	40	19.45	25	45.34	35	14.26
Jinhua	33	44.69	3	92.11	35	33.47	36	26.45	34	39.74	29	31.67
Taizhou	34	43.26	21	75.00	24	57.82	21	37.18	39	32.72	36	13.61
Cologne	35	43.16	32	50.00	40	21.93	26	34.18	15	53.39	12	56.32
Zhuhai	36	42.05	16	76.11	25	57.13	38	22.11	28	42.93	40	11.94
Lyon	37	41.70	30	53.09	36	32.01	33	29.22	22	48.01	26	46.19
Frederick Port	38	36.38	39	25.00	34	34.04	32	29.28	27	43.26	24	50.30
Malaga	39	34.89	36	37.50	41	18.25	31	30.74	11	56.35	30	31.62
Santande	40	32.41	33	50.00	37	28.87	39	21.71	41	31.02	31	30.47
Verona	41	25.81	41	12.50	39	24.40	41	12.27	16	52.63	33	27.26

The calculation principle in scoring is shown in the Affix III, IV and V. As for the domestic and international cities, different data sources make the different units, such as The online Publicity of the Government non Secret-related Official Documents, The Participation Ratio of the Public online, the proportion of R&D expenditure in GDP, The Coverage Density of the free network in the public space, The Proportion of the Internet Users, The Per Capita Net Purchase Expenditure and so on, and the data sources of the international cities shall prevail

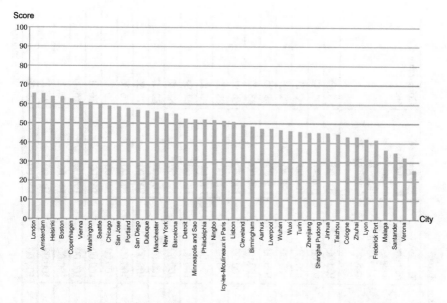

Fig. 7 The 41 cities' comprehensive score of intelligent construction at home and abroad

points), Boston, USA (63.87 points), Copenhagen, Denmark (62.92 points), Vienna, Austria (61.22 points), Washington Dc, United States (60.92 points) and Seattle, USA (60.02 points). Most of these cities are capitals which have gathered each country's a large number of economic and social resources, and also made the lead in the positioning and extension of the concept of intelligent cities.

Cities whose comprehensive scores are around 30 points include Verona, Italy (25.81 points), Santander, Spain (32.41 points), Malaga, Spain (34.89 points) and Fredrick port, Germany (36.38 points). These cities put more focus on one particular aspect of the construction of intelligent cities, such as Fredrick port, Germany, whose practice of intelligent cities takes the mode of the cooperation between government support and Germany telecommunications company, and mainly promotes the ICT application while being less involved in other ways. All these result in the lower comprehensive scores and uneven scores in five aspects.

The above 41 cities' Intelligent Environment and construction, Intelligent Management and Services, Intelligent Economy and Industries, Intelligent Hardware Facilities and Residents' Intelligent Literacy are shown as Tables 8, 9, 10, 11 and 12 and Figs. 8, 9, 10, 11 and 12.

In the matter of intelligent Environment and Construction (see Table 8 and Fig. 8), the highest score (97.84 points) that is owned by Amsterdam, Holland is nearly 7 times as many as the lowest score (12.50 points) that is owned by Verona, Italy. It's worth noting that these 41 cities have formed three distinct echelons in the aspect of intelligent environment and construction, 22 of which get the score over 70 points, 12 of which get the score around 50 points while 7 of which get the score less than 40 points. The reason causes this ladder-like distribution is that only one

Table 8 The evaluation of the 41 cities' intelligent environment and construction at home and abroad

Ranking	City	The density of city PM201/PM100 monitoring sites	The level of the coverage of city network management	The level of citizens' usage of intelligent traffic tools	The level of the publish online of the city's future development plan	Comprehensive score
1	Amsterdam	91.35	100.00	100.00	100.00	97.84
2	Seattle	69.72	100.00	100.00	100.00	92.43
3	Jinhua	68.45	100.00	100.00	100.00	92.11
4	Vienna	68.11	100.00	100.00	100.00	92.03
5	Zhenjiang	58.98	100.00	100.00	100.00	89.75
6	Boston	53.67	100.00	100.00	100.00	88.42
7	Ningbo	47.24	100.00	100.00	100.00	86.81
8	Copenhagen	43.59	100.00	100.00	100.00	85.90
9	Wuxi	40.74	100.00	100.00	100.00	85.18
10	Helsinki	39.36	100.00	100.00	100.00	84.84
11	Wuhan	29.90	100.00	100.00	100.00	82.48
12	Manchester	29.10	100.00	100.00	100.00	82.27
13	London	10.65	100.00	100.00	100.00	77.66
14	San Jose	7.43	100.00	100.00	100.00	76.86
15	Philadelphia	4.84	100.00	100.00	100.00	76.21
16	Zhuhai	54.43	50.00	100.00	100.00	76.11
17	San Diego	4.01	100.00	100.00	100.00	76.00
18	Chicago	2.80	100.00	100.00	100.00	75.70
19	Shanghai Pudong	2.78	100.00	100.00	100.00	75.70
20	Barcelona	0.00	100.00	100.00	100.00	75.00

(continued)

Table 8 (continued)

Ranking	City	The density of city PM201/PM100 monitoring sites	The level of the coverage of city network management	The level of citizens' usage of intelligent traffic tools	The level of the publish online of the city's future development plan	Comprehensive score
21	Taizhou	100.00	50.00	50.00	100.00	75.00
22	New York	38.38	100.00	100.00	50.00	72.10
23	Portland	24.18	100.00	100.00	50.00	68.55
24	Washington DC	21.16	100.00	100.00	50.00	67.79
25	San Francisco	16.83	100.00	100.00	50.00	66.71
26	Dubuque	0.00	50.00	100.00	100.00	62.50
27	Minneapolis and Sao Paulo	0.00	100.00	100.00	50.00	62.50
28	Lisbon	0.00	50.00	100.00	100.00	62.50
29	Aarhus	0.00	100.00	100.00	50.00	62.50
30	Lyon	12.34	100.00	0.00	100.00	53.09
31	Turin	0.00	0.00	100.00	100.00	50.00
32	Cologne	0.00	0.00	100.00	100.00	50.00
33	Santander	0.00	0.00	100.00	100.00	50.00
34	Detroit	28.09	100.00	0.00	50.00	44.52
35	Cleveland	0.00	0.00	100.00	50.00	37.50
36	Malaga	0.00	0.00	100.00	50.00	37.50
37	Birmingham	12.57	0.00	0.00	100.00	28.14
38	Issy-les-Moulineaux, Paris	0.00	0.00	0.00	100.00	25.00
39	Frederick port	0.00	0.00	0.00	100.00	25.00
40	Liverpool	30.09	0.00	0.00	50.00	20.02
41	Verona	0.00	0.00	0.00	50.00	12.50

Table 9 The evaluation of the 41 cities' intelligent management and services at home and abroad

Ranking	City	The online publicity of the government non secret-related official documents	The participation ratio of the public online	The level of the citizens' usage of the electronic health record	The level of the emergency intelligent reaction	Comprehensive score
1	Helsinki	97.80	2.13	100.00	100.00	74.98
2	Boston	80.22	17.23	100.00	100.00	74.36
3	Amsterdam	91.21	0.24	100.00	100.00	72.86
4	London	83.52	4.68	100.00	100.00	72.05
5	San Diego	80.22	6.18	100.00	100.00	71.60
6	Ningbo	39.57	46.51	100.00	100.00	71.52
7	Birmingham	83.52	1.28	100.00	100.00	71.20
8	Dubuque	80.22	3.59	100.00	100.00	70.95
9	Shanghai Pudong	78.58	3.67	100.00	100.00	70.56
10	Portland	80.22	0.67	100.00	100.00	70.22
11	San Jose	80.22	0.43	100.00	100.00	70.16
12	Detroit	80.22	0.37	100.00	100.00	70.15
13	Chicago	80.22	0.15	100.00	100.00	70.09
14	Wuhan	40.38	36.36	100.00	100.00	69.19
15	Vienna	75.82	0.09	100.00	100.00	68.98
16	Lisbon	68.13	0.06	100.00	100.00	67.05
17	Wuxi	38.04	63.50	50.00	100.00	62.89
18	Turin	47.25	0.11	100.00	100.00	61.84
19	Washington DC	80.22	15.95	100.00	50.00	61.54
20	Minneapolis and Sao Paulo	80.22	9.20	100.00	50.00	59.85

(continued)

Table 9 (continued)

Ranking	City	The online publicity of the government non secret-related official documents	The participation ratio of the public online	The level of the citizens' usage of the electronic health record	The level of the emergency intelligent reaction	Comprehensive score
21	Cleveland	80.22	6.09	100.00	50.00	59.08
22	Liverpool	83.52	0.13	50.00	100.00	58.41
23	Issy-les-Moulineaux, Paris	78.02	4.73	100.00	50.00	58.19
24	Taizhou	31.27	100.00	50.00	50.00	57.82
25	Zhuhai	71.57	6.97	50.00	100.00	57.13
26	Barcelona	64.84	0.02	100.00	50.00	53.71
27	Copenhagen	100.00	2.40	100.00	0.00	50.60
28	Manchester	83.52	8.62	0.00	100.00	48.04
29	Zhenjiang	37.36	52.00	100.00	0.00	47.34
30	San Francisco	80.22	8.27	0.00	100.00	47.12
31	Philadelphia	80.22	5.03	100.00	0.00	46.31
32	New York	80.22	3.59	50.00	50.00	45.95
33	Seattle	80.22	3.41	0.00	100.00	45.91
34	Frederick port	85.71	0.43	50.00	0.00	34.04
35	Jinhua	30.88	3.02	100.00	0.00	33.47
36	Lyon	78.02	0.04	50.00	0.00	32.01
37	Santander	64.84	0.63	50.00	0.00	28.87
38	Aarhus	100.00	0.27	0.00	0.00	25.07
39	Verona	47.25	0.34	50.00	0.00	24.40
40	Cologne	85.71	2.01	0.00	0.00	21.93
41	Malaga	64.84	8.16	0.00	0.00	18.25

Table 10 The evaluation of the 41 cities' intelligent economy and industries at home and abroad

Ranking	City	The proportion of R&D expenditure in GDP	Urban labor productivity	City output density	Proportion of urban intelligent industry	Comprehensive score
1	Washington DC	78.59	100.00	100.00	23.72	75.58
2	Liverpool	48.45	45.01	51.39	100.00	61.21
3	Boston	78.59	77.30	58.90	23.72	59.63
4	Copenhagen	87.32	31.85	57.61	61.60	59.60
5	Amsterdam	60.85	61.68	62.65	31.40	54.14
6	London	48.45	13.53	51.36	100.00	53.33
7	Manchester	48.45	27.34	32.55	100.00	52.08
8	Helsinki	100.00	19.14	4.50	66.55	47.55
9	Seattle	78.59	59.24	26.74	23.72	47.07
10	Barcelona	84.51	16.42	71.54	14.33	46.70
11	Aarhus	87.32	19.22	15.02	61.60	45.79
12	Birmingham	48.45	16.29	18.20	100.00	45.73
13	San Francisco	78.59	57.00	21.76	23.72	45.27
14	New York	78.59	22.60	51.58	23.72	44.12
15	Minneapolis and Sao Paulo	78.59	44.77	27.58	23.72	43.67
16	Philadelphia	78.59	32.50	37.66	23.72	43.12
17	Chicago	78.59	30.30	36.90	23.72	42.38
18	Detroit	78.59	42.13	21.23	23.72	41.42
19	Cleveland	78.59	39.63	20.16	23.72	40.52
20	Portland	78.59	37.15	15.80	23.72	38.82
21	Taizhou	55.77	8.21	20.65	64.08	37.18
22	Vienna	67.32	15.85	18.49	41.64	35.83

(continued)

Table 10 (continued)

Ranking	City	The proportion of R&D expenditure in GDP	Urban labor productivity	City output density	Proportion of urban intelligent industry	Comprehensive score
23	Wuxi	55.77	7.54	13.83	64.08	35.31
24	San Jose	78.59	22.64	13.31	23.72	34.57
25	Zhenjiang	55.77	6.29	11.90	64.08	34.51
26	Cologne	82.25	16.46	11.39	26.62	34.18
27	Turin	84.51	13.10	25.12	13.14	33.97
28	San Diego	78.59	21.79	8.33	23.72	33.11
29	Lisbon	42.25	27.07	47.86	10.41	31.90
30	Issy-les-Moulineaux, Paris	63.66	10.01	10.03	40.10	30.95
31	Malaga	36.62	51.65	20.35	14.33	30.74
32	Frederick port	82.25	6.69	1.55	26.62	29.28
33	Lyon	63.66	9.07	4.04	40.10	29.22
34	Dubuque	78.59	6.21	1.21	23.72	27.43
35	Wuhan	55.77	2.52	7.56	42.40	27.06
36	Jinhua	55.77	11.44	19.72	18.87	26.45
37	Ningbo	55.77	4.88	10.55	18.94	22.54
38	Zhuhai	55.77	4.08	6.75	21.84	22.11
39	Santander	36.62	14.98	20.91	14.33	21.71
40	Shanghai Pudong	55.77	2.69	3.15	16.21	19.45
41	Verona	35.77	0.12	0.04	13.14	12.27

Table 11 The evaluation of the 41 cities' intelligent hardware facilities at home and abroad

Ranking	City	The coverage density of the free network in the public space	Per capita usage of the mobile network	Urban broadband speed	The level of the coverage of the smart grid	Comprehensive score
1	Issy-les-Moulineaux, Paris	17.78	67.31	100.00	100.00	71.27
2	Barcelona	39.45	92.69	42.77	100.00	68.73
3	Ningbo	100.00	33.46	26.55	100.00	65.00
4	London	33.56	92.95	27.38	100.00	63.47
5	Copenhagen	17.18	86.61	47.10	100.00	62.72
6	Lisbon	17.29	93.12	39.31	100.00	62.43
7	Vienna	10.20	91.87	41.81	100.00	60.97
8	New York	22.00	68.25	53.06	100.00	60.83
9	Seattle	21.77	68.25	48.09	100.00	59.53
10	Amsterdam	6.13	77.36	42.74	100.00	56.56
11	Malaga	0.19	92.69	32.53	100.00	56.35
12	Portland	16.13	68.25	34.81	100.00	54.80
13	Turin	2.21	100.00	12.19	100.00	53.60
14	San Jose	6.84	68.25	39.21	100.00	53.58
15	Cologne	2.28	83.22	28.05	100.00	53.39
16	Verona	0.29	100.00	10.21	100.00	52.63
17	Chicago	10.22	68.25	28.81	100.00	51.82
18	Dubuque	1.31	68.25	31.20	100.00	50.19
19	Detroit	6.90	68.25	24.71	100.00	49.97
20	Cleveland	9.62	68.25	21.47	100.00	49.84
21	San Diego	3.42	68.25	27.10	100.00	49.69

(continued)

Table 11 (continued)

Ranking	City	The coverage density of the free network in the public space	Per capita usage of the mobile network	Urban broadband speed	The level of the coverage of the smart grid	Comprehensive score
22	Lyon	1.66	67.31	73.06	50.00	48.01
23	Helsinki	2.15	87.92	50.03	50.00	47.53
24	Washington DC	30.22	68.25	34.92	50.00	45.85
25	Shanghai Pudong	12.49	33.46	35.40	100.00	45.34
26	Aarhus	2.49	86.61	41.91	50.00	45.25
27	Frederick port	3.24	83.22	36.59	50.00	43.26
28	Zhuhai	9.55	33.46	28.72	100.00	42.93
29	Zhenjiang	15.53	33.46	21.06	100.00	42.51
30	Wuhan	9.09	33.46	24.87	100.00	41.86
31	Boston	13.80	68.25	33.11	50.00	41.29
32	San Francisco	9.26	68.25	36.07	50.00	40.89
33	Philadelphia	7.33	68.25	35.78	50.00	40.34
34	Jinhua	47.17	33.46	28.31	50.00	39.74
35	Minneapolis and Sao Paulo	14.92	68.25	23.47	50.00	39.16
36	Manchester	21.17	92.95	26.24	0.00	35.09
37	Liverpool	13.38	92.95	30.05	0.00	34.10
38	Wuxi	16.22	33.46	33.58	50.00	33.31
39	Taizhou	18.62	33.46	28.78	50.00	32.72
40	Birmingham	5.59	92.95	31.49	0.00	32.51
41	Santander	6.48	92.69	24.93	0.00	31.02

Table 12 The evaluation of the 41 cities' residents' intelligent literacy at home and abroad

Ranking	City	The proportion of the internet users	The proportion of information practitioners	The proportion of the population with college level or above	The per capita net purchase expenditure	Comprehensive score
1	Dubuque	47.45	75.23	100.00	65.47	72.04
2	Issy-les-Moulineaux, Paris	55.10	100.00	80.18	50.92	71.55
3	Helsinki	56.02	10.85	93.73	100.00	65.15
4	Manchester	54.98	17.98	86.00	95.38	63.59
5	London	54.98	11.03	86.00	95.38	61.85
6	Birmingham	54.98	2.81	86.00	95.38	59.79
7	Liverpool	54.98	1.74	86.00	95.38	59.53
8	San Jose	48.80	20.44	100.00	65.47	58.67
9	Aarhus	57.92	13.92	86.11	75.65	58.40
10	Portland	52.72	10.67	100.00	65.47	57.21
11	Detroit	48.00	12.48	100.00	65.47	56.49
12	Cologne	51.43	32.91	96.64	44.31	56.32
13	Copenhagen	58.16	3.18	86.11	75.65	55.78
14	Boston	52.78	4.34	100.00	65.47	55.65
15	Minneapolis and Sao Paulo	50.27	6.69	100.00	65.47	55.61
16	Cleveland	46.96	9.04	100.00	65.47	55.37
17	Chicago	48.06	7.23	100.00	65.47	55.19
18	Seattle	52.47	2.71	100.00	65.47	55.16
19	San Diego	48.80	5.97	100.00	65.47	55.06
20	San Francisco	48.80	5.06	100.00	65.47	54.83

(continued)

Table 12 (continued)

Ranking	City	The proportion of the internet users	The proportion of information practitioners	The proportion of the population with college level or above	The per capita net purchase expenditure	Comprehensive score
21	Philadelphia	47.63	5.43	100.00	65.47	54.63
22	New York	49.90	2.89	100.00	65.47	54.56
23	Washington DC	46.84	3.07	100.00	65.47	53.84
24	Frederick port	60.25	0.00	96.64	44.31	50.30
25	Vienna	49.35	7.17	92.39	44.31	48.30
26	Lyon	50.14	3.50	80.18	50.92	46.19
27	Amsterdam	57.55	7.89	80.96	38.11	46.13
28	Barcelona	45.92	3.96	60.47	17.56	31.98
29	Jinhua	100.00	2.30	18.20	6.18	31.67
30	Malaga	43.84	4.61	60.47	17.56	31.62
31	Santander	43.84	0.00	60.47	17.56	30.47
32	Turin	35.82	0.00	62.71	10.52	27.26
33	Verona	35.82	0.00	62.71	10.52	27.26
34	Lisbon	38.02	0.00	39.19	17.56	23.69
35	Shanghai Pudong	47.91	0.51	2.44	6.18	14.26
36	Taizhou	38.76	1.56	7.94	6.18	13.61
37	Ningbo	40.13	0.72	6.73	6.18	13.44
38	Wuxi	37.00	0.61	6.52	6.18	12.58
39	Zhenjiang	28.77	0.54	12.30	6.18	11.95
40	Zhuhai	24.39	1.99	15.21	6.18	11.94
41	Wuhan	18.32	0.72	17.09	6.18	10.58

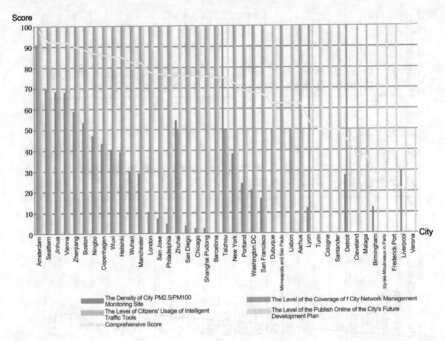

Fig. 8 The score of the 41 cities' intelligent environment and construction at home and abroad

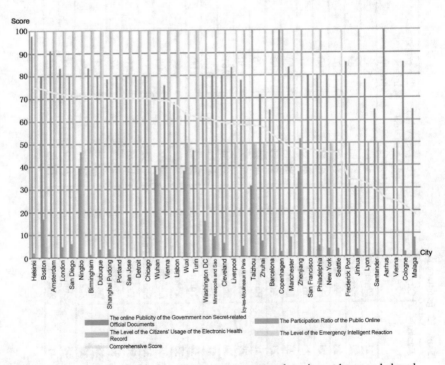

Fig. 9 The score of the 41 cities' intelligent management and services at home and abroad

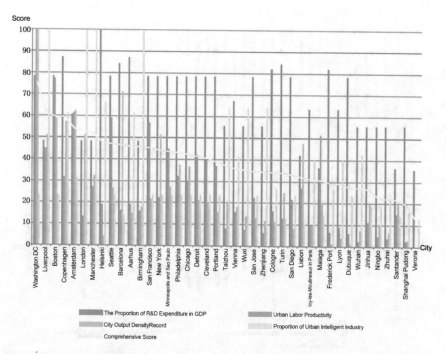

Fig. 10 The score of the 41 cities' intelligent economy and industries at home and abroad

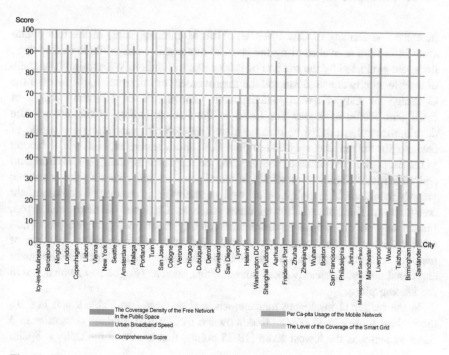

Fig. 11 The score of the 41 cities' intelligent hardware facilities at home and abroad

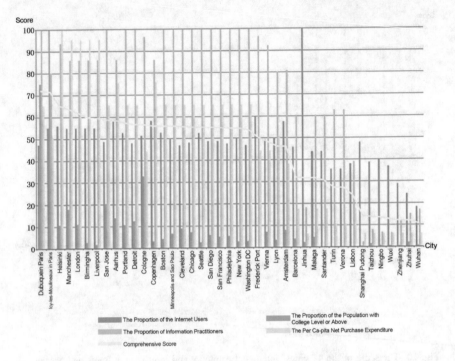

Fig. 12 The score of the 41 cities' residents' intelligent literacy at home and abroad

(namely The Density of City PM201/PM100 Monitoring Sites) of the four evaluation indicators is evaluated by the specific numerical value while the other three items are evaluated by the method of ladder-like evaluation. With the advancement of intelligent city construction and the improvement of information, the level of the Coverage of City Network Management, The Level of Citizens' Usage of Intelligent Traffic tools and The Level of the Publish Online of the city's future development plan can all be measured step by step by quantitative evaluation of data. As for the ranking, such as London, U.K. and Manchester, U.K. which belong to the industrial city, make the outstanding performance in intelligent environment and construction because of efforts in transformation and changing the economic growth model of high pollution as well as their big investment in environment construction. But some industrial cities, such as Cologne, Germany and Lyon, France and so on, whose environmental quality have no significant proportion relationship with construction investment. And as for Chinese cities, they are of medium level or slightly above generally in the aspect of intelligent environment and construction. Especially some coastal cities, such as Ningbo and Zhuhai, are in the leading place.

In the matter of intelligent management and services (see Table 9 and Fig. 9), the highest score (74.98 points) that is owned by Helsinki, Finland is more than 3 times as many as the lowest score (18.25 points) that is owned by Malaga, Spain.

It's easy to see that most of the leading cities are capitals or economic centers, such as Boston, USA, Amsterdam, Holland, London, U.K. and Shanghai Pudong, China and so on; their scores are all above 70 points, which have the connection with the policy guidance of the four indicators. But the cities with less than 40 points mostly focus on one aspect of intelligent city construction (such as ICT promotion), and ignore the aspect of management and services (such as public security and medical services), these typical cities include Jinhua, China, Cologne, Germany, Lyon, France and Frederic port, Germany and so on. In the matter of intelligent management and services, Chinese cities are ranked in either leading place or lowing place, for example, Ningbo, Shanghai Pudong and Wuhan are in the front rank while Jinhua and Zhenjiang are of the low ranked locations, and this ranking has the significant connection with the city level and policy tendency.

In the matter of intelligent economy and industries (see Table 10 and Fig. 10), the highest score (75.58 points) owned by Washington DC, USA is 5 times as many as the lowest score (12.27 points) that is owned by Verona, Italy. Most of the capitals are ranked highly with the score above 50 points, such as Amsterdam, Holland, Copenhagen, Denmark and London, U.K. It's easy to see that there is no significant difference among economy and industries of international intelligent cities. But in China, the cities which do well in this aspect are Taizhou, Wuxi and Zhenjiang and so on, and they are ranked at front in the selected cities both in intelligent industries and scientific budgets.

In the matter of intelligent hardware facilities (see Table 11 and Fig. 11), the highest score (71.27 points) that is owned by Issy-les-Moulineaux, Paris is nearly 1.3 times the lowest score (31.02 points) that is owned by Santander, Spain. London, U.K., Copenhagen, Denmark, and Lisbon, Portugal all have the scores over 60 points. This shows that many countries' capitals are the gathering place of all aspects of resources, and they also put large investment in hardware facilities with effective results. But in China, because of the different characteristic construction, there are significant differences between fund investment and effective construction in some cities, and Ningbo ranks at the third while Shanghai Pudong, Zhuhai, Zhenjiang and Wuhan rank at the middle.

In the matter of Residents' Intelligent Literacy (see Table 12 and Fig. 12), the highest score (72.04 points) that is owned by Dubuque, USA is about 6 times the lowest score (10.58 points) that is owned by Wuhan, China. The capital city doesn't show any obvious advantages, and because of the high ranking of the Proportion of the Internet Users, the proportion of the population with College level or above and The Per Capita Net Purchase Expenditure in England, Manchester, London, Birmingham and Liverpool are all ranked highly with the score around 60 points. But most of Chinese cities do not do well in ranking as 7 cities' scores are less than 15 points in the total 8 cities.

3   Growth and Development Stage Analysis of Global Intelligent Cities

The process of a city's constant development and construction can take the forms of a variety of phenomena, such as expansion, gathering, shrinking, mutation and dying and so on. A city's development is not only the growth of amount and the expansion of the scale, but also the qualitative changes, such as industrial upgrading and function transition, to gradually form the complete function and self-organization construction. A city's development is a progress from disorder to order, and shows the nonlinear, self-organization and spiral upward development features. At the same time, a city's development is controlled by its own conditions and self-organization principles, and it constantly adjusts and improves itself during the energy exchange with the outside world to keep moving forward and upward direction.

The growth and development of intelligent cities include two dimensions, one is the city's own growth and development as an organism, the other is advanced technology guiding the city to grow and develop toward the direction of Intelligence. These two dimensions are complementary and mutual influencing. The self-growth and development level of a city is the basis of its intelligent ones while the intelligence brought by the technology brings more and better opportunities and conditions for the city itself.

The city's growth level includes three parts namely intelligent environment and construction, intelligent economy and industries and intelligent hardware facilities which have connection with a city's growth stated in the preceding intelligent city evaluation. And the development level is the average value of the intelligent management and services and Residents' Intelligent literacy, as Figs. 13 and 14.

In Fig. 14, 50 is the high-low demarcation of growth and development level, and the 41 cities at home and abroad are put into four quadrants, which we are going to analyze the situation of intelligent cities' growth and development in each quadrant according to as follows.

3.1   Low Growth, Low Development

The cities at this stage include most of the Chinese intelligent cities and some European intelligent cities, such as Shanghai Pudong (see Fig. 15), Zhuhai, Wuhan and Taizhou from China, Aarhus, Denmark, Cologne and Frederick port from Germany, Lyon, France, Santander and Malaga from Spain, Lisbon, Portugal and Verona and Turin from Italy.

Most of Chinese intelligent cities are at this stage with the feature of low growth and low development, which is closely related to the differences in city basis between China and Europe and the United States. But the cities which are at the stage in the European intelligent cities are the cities which choose one particular

Development level

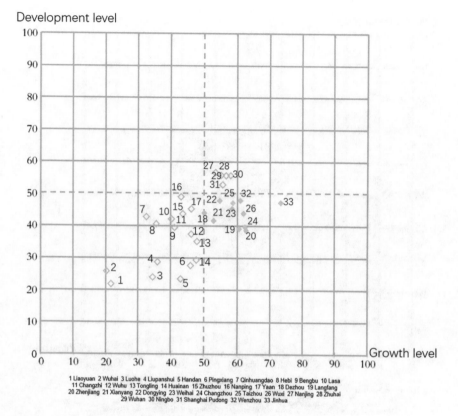

1 Liaoyuan 2 Wuhai 3 Luohe 4 Liupanshui 5 Handan 6 Pingxiang 7 Qinhuangdao 8 Hebi 9 Bengbu 10 Lasa
11 Changzhi 12 Wuhu 13 Tongling 14 Huainan 15 Zhuzhou 16 Nanping 17 Yaan 18 Dezhou 19 Langfang
20 Zhenjiang 21 Xianyang 22 Dongying 23 Weihai 24 Changzhou 25 Taizhou 26 Wuxi 27 Nanjing 28 Zhuhai
29 Wuhan 30 Ningbo 31 Shanghai Pudong 32 Wenzhou 33 Jinhua

Fig. 13 The domestic 33 cities' growth and development level analysis

aspect to construct. And most of the cities at this stage only make the "intelligent cities" as a slogan, while the level of actual intelligent city construction is only at the primary stage, and the government puts more focus on the other industries' development with intelligent cities as gimmicks, instead of the serious construction of intelligent cities.

3.2   High Growth, Low Development

The cities at this stage include Barcelona, Spain and Chinese Ningbo, Jinhua, Wuxi and Zhenjiang. Barcelona, as an important city in Spain which is also a major country in southern Europe, makes a lot of attempts in intelligent city construction and also puts forward higher requirements. From the outset, Barcelona implements the concept of intelligent city throughout the process of city construction, representing a future trend, but due to the long construction of a new city, at present it has not yet reached a better result.

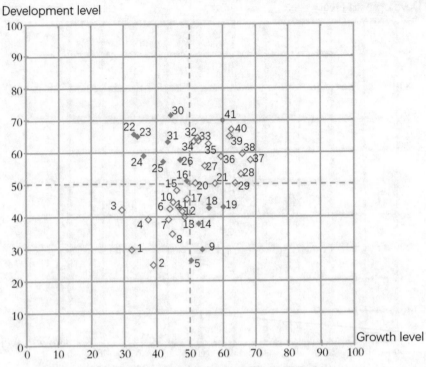

1 Santander 2 Malaga 3 Frederick Port 4 Lyon 5 Jinhua 6 Shanghai Pudong 7 Cologne 8 Zhuhai 9 Zhenjiang
10 Turin 11 Verona 12 Aarhus 13 Wuhan 14 Wuxi 15 Taizhou 16 San Francisco 17 Lisbon 18 Ningbo 19 Barcelona 20 Philadelphia 21
New York 22 Birmingham 23 Icy-les-Moulineaux in Paris 24 Liverpool 25 Cleveland
26 Minneapolis and Sao Paulo 27 Manchester 28 Copenhagen 29 Seattle 30 Dubuque 31 Detroit 32 San Jose
33 Portland 34 San Diego 35 Chicago 36 Vienna 37 Washington DC 38 Amsterdam 39 Boston 40 London 41 Helsinki

Fig. 14 The 41 cities' growth and development level analysis of intelligent cities in the world

Due to the preference of policy setting and investment effort, it is made in the advanced level in intelligent environment and construction, intelligent economy and industries, and intelligent hardware facilities, and thereby has a higher level of intelligent growth (see Fig. 16). However, intelligent management and services and Residents' Intelligent literacy which relate to the intelligent development level wouldn't make an obvious progress in a short period. It also reflects the focus of the next phase of intelligent city construction in Ningbo.

3.3   Low Growth, High Development

The cities at this stage include Issy-les-Moulineaux, Paris, France, Dubuque, Detroit, Cleveland Minneapolis and Sao Paulo from America and the cities of Birmingham and Liverpool form England.

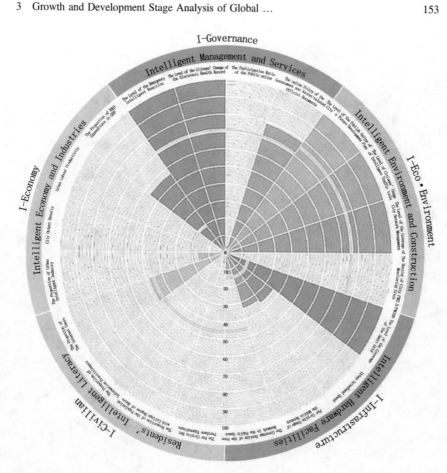

Fig. 15 The intelligent city evaluation of Shanghai Pudong

Take the city of Issy-les-Moulineaux, Paris from France as the example. The local government takes the optimization of public services as the entry point of promoting the intelligent city construction, and adopts more pragmatic practical actions which refer to intelligent energy, intelligent business and intelligent education, thus it achieves a higher level of intelligent development (Fig. 17).

3.4 High Growth, High Development

The cities at this stage include London and Manchester from England, Amsterdam from Holland, Helsinki from Finland, Vienna from Austria, Copenhagen from Denmark, as well as Boston, Washington DC, Chicago, Portland, San Jose, San Diego, Seattle, New York and Philadelphia from America.

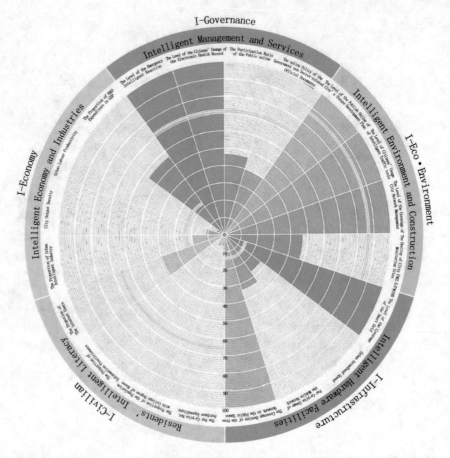

Fig. 16 The intelligent city evaluation of Ningbo

Taking Amsterdam as the example, "the Wisdom City of Amsterdam" is a cooperative plan of business, government and communities, its goal is to save energy and reduce carbon dioxides by sustainable working, living, transportation and public spaces. Amsterdam has planned by 2025 to reduce carbon dioxides by 40% compared with 1990, and expects to achieve this by 2015 in advance with the implementation of the project of intelligent cites. At the same time, the project puts forward 4 target realms, including sustainable living, work, transportation and places.

Although Amsterdam's goal of building wisdom cities is simple, which only focuses on the technology applications related with energy saving, yet it does have a strong maneuverability and can be of a great benefit to be taken into action immediately and build a city's brand of wisdom. Under the goal of energy conservation and carbon reduction, there are still several systems, and by focusing on the integration of the various systems, all these make the high-level development

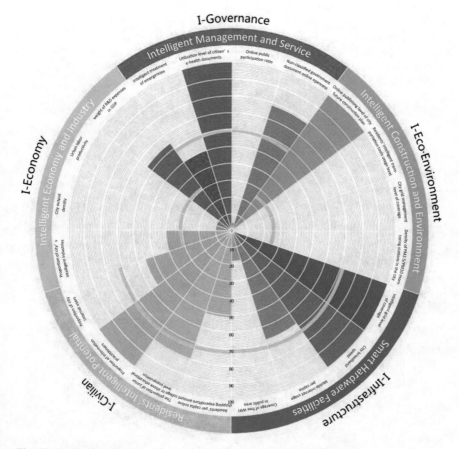

Fig. 17 The intelligent city evaluation of Issy-les-Moulineaux, Paris, France

come true. At the same time, there are also a lot of scientific researches and government collaboration which help to achieve the high-level development. All in all, Amsterdam's intelligent city construction is at the leading place in the world (Fig. 18).

4 Construction Trends of Global Intelligent Cities

4.1 Global In-Depth Cooperation

Since 2008, intelligent or wisdom city has been the hot keyword of city construction, and many cities have practiced this and several Intelligent or Wisdom City Forums haven been held at home and abroad.

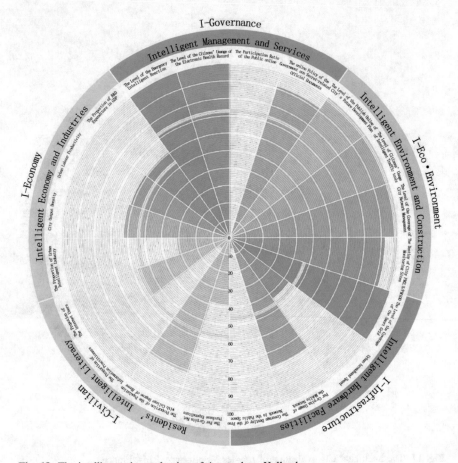

Fig. 18 The intelligent city evaluation of Amsterdam, Holland

In China, the Product Conference on Technology and Application of Wisdom City in Ningbo (see Fig. 19), is the first State-level conference with the theme of Wisdom City. It's jointly organized by 8 institutions, including Ministry of Industry and Information Technology, The State Press and Publication Administration, The Chinese Academy of Sciences, the Chinese Academy of Engineering, China Telecom, China Mobile, China Unicom and the People's Government of Ningbo Municipality, and it's also a professional exhibition of new technologies, new achievements and new products on wisdom city construction at home and abroad.

At abroad, the Wisdom City Conference in Barcelona (see Fig. 20), as the largest global platform of wisdom city on exhibition activities and research, brings together worldwide representative cities, top manufactures, solutions, experts and scholars and well-known medias, and provides a complete panoramic view for the world wisdom city construction and development.

Fig. 19 The product conference on technology and application of Wisdom City in Ningbo, China. *Photo Credit* http://www.expo-nb.com/content.aspx?id=1373

Fig. 20 The wisdom city conference in Barcelona. *Photo Credit* http://www.hotelactual.com/content/en/agenda/smart-city-barcelona-2014

Fig. 21   EU-China innovation cooperation dialogue. *Photo Credit* http://www.most.gov.cn/kjbgz/201311/t20131125_110569.htm

Since EU-China Innovation Cooperation Dialogue held in Beijing in November 2013, EU-China exchanges at the city level has been gradually carried out, and a number of pilot cities have been selected, which officially launched the EU-China intelligent city cooperation (see Fig. 21). In addition, the Sino-German Forum on intelligent city which is organized by intelligent city research community in Munich, aims at making the cities in China and Germany to cooperate in the field of infrastructure, business model, the government's role and evaluation systems.

It's easy to see from the intelligent city forum's present situations and scale of impact at home and abroad, that the national forum plays a very important role in guiding the national construction of intelligent cities. However, domestic and foreign exchanges only stay in the city project level, such as the exchange of designation and concept, and lack the research institutions, academia and folk exchange.

The next step is to strengthen the world in-depth cooperation in the field of intelligent city construction. And the relevant research institutions should establish the long-term exchanges and visits and regular meeting mechanism, build the global intelligent network for urban development, and train the professional and technical talents to support city construction, promotion and improvement, in order to accelerate the intelligent cities' development and progress.

4.2   *From Hardware to Social Quality*

At the beginning of wisdom city concept, most of the cities especially paid much attention to the Information-based infrastructure construction and the new technology's application in various city systems, as well as more focused on the inputs rather than outputs. In fact, the aim of the wisdom city is to achieve a city's overall harmony and sustainability, and the tenet is to serve the aim by taking information and new technology as a tool and approach. At the situation of being able to achieve the corresponding goal, simple low-tech solutions should be the first choice. If technology development is emphasized unilaterally in intelligent city construction, there would be the accumulation of many technical problems.

4.3   *From a Single City to City Agglomeration Area*

The construction development of intelligent cities will lead to the formation of town groups, which may be the ultimate meaning of the development of intelligent cities. At present, competitions between countries has grown from a single city to the town clusters which unite surrounding multilevel cities with key cities as the core. Regardless of environment, transportation, economy, society or culture, compared with single cities, town cluster put forward the higher demands on development and construction, such as more integrated and collaborative regional resources, regional ecological security, the accurate links between regional fast-track and the traffic in the city, regional industry layout, and the regional cultural identity and so on. The practice from an intelligent city to intelligent town cluster provides a more rational and effective way for improving the national competitiveness, but up to now, this hasn't been covered in the construction of intelligent city at home and abroad.

Chapter 9
Promoting Strategy of Intelligent City Evaluation Indicator System

Abstract The final chapter provides the promoting strategies of intelligent city evaluation indicator system from the perspectives of international cooperation, updating and reporting mechanism, implementation and upgrading, and public participation.

Keywords Intelligent city · Evaluation indicator system · Recommendation Promoting strategies

1 Cooperation with Europe and Explosion to the International Environment to Form an Authoritative Third-Party Evaluation Indicator System

This group has made consensus on the cooperation with international well-known authoritative research institutions, such as Chinese Academy of Engineering (CAE), German Academy of Engineering, Sweden Royal Academy of Engineering and Holland Academy of Engineering, to strengthen the corporation of intelligent city evaluation indicator system. The aim is to realize mutual learning of the intelligent city evaluation systems on the platform of central Europe's coordinated development of urbanization, to establish a widely acknowledged third party evaluation indicator system with authority in China and Europe, and meanwhile to promote the continuous development and evolution of this evaluation indicator system by practice of evaluation of the intelligent city evaluation system in the multiple backgrounds of China and Europe.

© Springer Nature Singapore Pte Ltd. and Zhejiang University Press 2018   161
Z. Wu, *Intelligent City Evaluation System*, Strategic Research
on Construction and Promotion of China's Intelligent Cities,
https://doi.org/10.1007/978-981-10-5939-1_9

2   Cooperation with Asia and America to Develop Academic Cooperation Among Professional Institutions

This is to promote the cooperation between China and other Asian countries and American countries at the same time, especially strengthen the communication on academic with main developed countries, such as South Korea, Singapore, the United States, establish a long term exchange of visits and regular meeting mechanism, build a developmental network for intelligent city globally-covered, absorb professionals in order to support the further promotion and improvement of the intelligent city evaluation system, and promote the development and progress of the intelligent city itself.

3   Prompt Release of 2015 Intelligent City Evaluation Indicator System, with an Annual Updating and Reporting Mechanism to Perfect It Within Practice

Results of Intelligent City Evaluation Indicator System of this research group has been regarded as a trial version, continuing to be deepened and improved on the basis of the existing pilot cities, to form the follow-up results including "evaluation system", "evaluation update system", "evaluation promotion strategy", to publish the new version of "Intelligent City Evaluation Indicator System Research", and to form practicality promotion and guidance suggestions aimed at the participating cities and alternative cities.

More and more cities put forward the application and evaluation of Intelligent City, which can continue to promote the updating and improvement of evaluation indicator system. With the actual development of the city and the overall level of development of the target city group as the basis for the indicator's revision and update we perfected practices and made constant upgrading in the practice of Intelligent City Evaluation Indicator System.

4   Real-Time Learning and Dynamic Evaluation and Implementation of Tracking and Diagnosis of Cities Claiming to Be Intellectualized

National intelligent city data bank shall be established to completely cover all the urban areas of the intelligent city for the purpose of the construction, catch the newly added urban areas in time, establish early warning mechanism to the non-intelligent cities, track and evaluate dynamically all of the cities in bank, including the independent evaluation and overall evaluation of community, regularly publish the evaluation results, release a real-time intelligent city rankings, put forward suggestions on the construction of city intelligent construction diagnosis, enhance the instrumentality and universality to the evaluation indicator system, and continue to update accumulated basic information and practical experience of the evaluation indicator system.

5   Strengthened Cooperation with Local Governments to Push the Implementation, Promotion, Feedback and Upgrading of Intelligent City Evaluation Indicator System

The interaction with the local government shall be strengthened to achieve the promotion agreement, build the normal communication mechanism between the local government and professional evaluation agencies, establish a more frequent and stable evaluation system mechanism on the basis of the evaluated cities that participate in the evaluation indicator system.

It is necessary to find and cope the weaknesses and shortcomings of the intelligent city construction, revise and adjust the implementation policies in time according to the evaluation indicator system and meanwhile through the evaluation indicator system of the local government's promotion, to use the intelligent city evaluation indicator system to obtain more reliable data sources, and to access to the implementation feedback of the evaluation indicator system and overall improvement.

6   Establishing Online Citizen Evaluation Platform to Get Authentic Feedback from the Public and Realizing the Monitoring Effect and Bottom-Up Public Participation

The intelligent city evaluation indicator system needs to come from people, and also needs to really serve people, emphasizing on the real feelings from the public to form a feedback mechanism for the evaluation process, evaluation results, public the calculation methods and ranking results of the evaluation indicator system online, and set up a network of information exchange platform and network supervision channels between professional analysis institutions and citizens. It needs to fully absorb public opinion, to form the bottom-up feedback mechanism of the intelligent city evaluation indicator system.

At the same time, the propaganda of the concept of intelligent city needs to be strengthened to raise public awareness and encourage judgement of the intelligent city, and promote the quality of intelligent residents through the promotion of evaluation indicator system. Professional assessment agencies can also form an analysis report based on public participation, public opinion feedback and other information, which are the results of the annual intelligent city evaluation indicator system.

Appendix
A.1 An Overview on Intelligent City Evaluation Systems in the World

See Tables A.1, A.2, A.3, A.4, A.5, A.6, A.7, A.8, A.9, A.10, A.11, A.12, A.13 and A.14.

© Springer Nature Singapore Pte Ltd. and Zhejiang University Press 2018
Z. Wu, *Intelligent City Evaluation System*, Strategic Research
on Construction and Promotion of China's Intelligent Cities,
https://doi.org/10.1007/978-981-10-5939-1

Table A.1 National intelligent city (district, town) pilot indicator system

First grade indicator	Secondary indicators	Tertiary indicators	Indicator description
Security system and infrastructure	Security system	Development plan outline and implementation plan for intelligent city	Refers to the integrity and feasibility of intelligent city development plan outline and implementation plan
		Organizations	Refers to the establishment of a special leadership organization system and the executive body, responsible for the creation of smart city
		Policies and regulations	Refers to the policies and regulations to ensure the construction and operation of intelligent city
		Budget planning and continuous protection	Refers to the budget planning and protection measures of the intelligent city's construction
		Operation management	Refers to the operation subject to definite the intelligent city and build the operated supervision system
	Network infrastructure	Wireless network	Refers to the basic conditions of wireless network coverage, speed
		Broadband network	Refers to the basic conditions of fixed broadband access coverage, access speed, including optical fiber
		Next generation broadcasting	Refers to the construction and usage of the next generation broadcasting
	Public platform and database	City public basic database	Refers to establish public basic databases, such as city basic spatial database, population basic database, basic database of legal person, macroeconomic database, building foundation database and etc.
		City public information platform	Refers to the construction of information platform to the city's all kinds of public information, which can be unified management and exchanged, meet the city all kinds of business and industry demand for public information exchange and service

(continued)

Table A.1 (continued)

First grade indicator	Secondary indicators	Tertiary indicators	Indicator description
Smart construction and livable	City construction administration	Information security	Refers to the safeguard measures and effectiveness of the intelligent city's information security
		Urban and rural planning	Refers to the preparation of complete and reasonable urban and rural planning, and according to the needs of the development of the city, making the road traffic planning, historical and cultural city protection planning, landscape planning and specific planning, to comprehensively guide the construction of city
		Digital city management	Refers to build a digital city management system based on the relevant national standards with city's geospatial framework, establish perfected assessment and incentive mechanism, to realize regional grid management
		Constructing market management	Refers to formulate the laws and regulations of constructing market management, and promote the government's abilities of supervision and management in the construction survey, design, construction and supervision by using the information means
		Estate management	Refers to promote the government's comprehensive management services abilities in a number of areas of housing planning, real estate sales, intermediary services, property mapping by formulating and implementing effective polices of real estate management, and use the information technology means to manage the real estate
Smart construction and livable	City construction management	Landscaping	Refers to promote the level of monitoring and management of landscaping, promote the level of urban garden greening through the application of advanced technology of remote sensing
		Protection in history and culture	Refers to promote the level of protection in urban history and culture through the application of information technology means
		Building energy supervision	Refers to promote the city's working level in building energy efficiency supervision, evaluation, control and management through the application of information technology means

(continued)

Table A.1 (continued)

First grade indicator	Secondary indicators	Tertiary indicators	Indicator description
	City function promotion	Green building	Refers to promote city's level in green building construction, management and evaluation through formulating effective policies and combining with the application of information technology means
		Water supply system	Refers to realize real-time monitoring and control to the whole water supply process from water source monitoring to tap water management, formulate reasonable information publicity system and guarantee the residential water security by using information technology means
		Drainage system	Refers to the construction of drainage facilities in discharging of life, industrial sewage, urban rainwater collection and channel, and promote the development of its overall function by using information technology means
		Water-saving application	Refers to the usage of city water-saving equipment and the cyclic utilization of water resources, and promote the overall level of its development by using information technology means
		Gas system	Refers to the popularity of city using clean, and promote the development of safe operation level by using information technology means
		Garbage classification and handling	Refers to the popularity of waste classification for the community and the handling ability of garbage harmless, and promote the overall level of its development by using information technology means
		Heating supply system	Refers to the northern city's winter heating facilities construction, promote the overall level of its development by using information technology means
		Lighting system	Refers to the coverage and energy-saving automation applied degree of city's all kinds if lightings
		Integrated management of underground pipe and space	Refers to realize urban underground pipe network digital integrated management and monitoring, and promote management level by using the technology means of 3D visualization

(continued)

Table A.1 (continued)

First grade indicator	Secondary indicators	Tertiary indicators	Indicator description
Smart management and service	Government affairs service	Decision support	Refers to information means and system supporting government decision-making
		Information discourse	Refers to public the government information in the field of the approval and implementation of budget balance, major construction projects, construction of public undertakings actively, timely and accurately through the government websites
		Online-do	Refers to perfect the function of the government portal website, expand the scope of Internet business, improve the efficiency of online business
Smart management and service	Government affairs service	Government affairs service system	Refers to the connection and integration of all types of government affairs service platforms, build up and linkage, clear hierarchy and government affairs service system of upper and lower linkage, clear hierarchy and covering urban and rural areas
	Basic public service	Basic public education	Refers to make a reasonable education development plan, use the information technology to improve target population, which can access to convenient level of basic public education services, and promote coverage and share of education resources
		Employment service	Refers to through the continuous improvement of the regulations and system, combining with the application of modern information technology, enhance the management level of city employment service, through the measures of the establishment of employment information service platform to promote the release of employment information ability, strengthen the guarantee of free employment training, protect the legitimate rights and interests of workers
		Social insurance	Refers to through the application of information technology, improve the target population to enjoy the basic old-age insurance, basic medical insurance, unemployment, work injury and maternity insurance service convenience degree, improve the quality supervision of social insurance services, improve the level of residents' life safeguard through the information service terminal construction on the basis of improving coverage scale
		Social services	Refers to through the application of information technology, improve the target population to enjoy the convenience degree of social relief, social welfare, the basic pension service and the entitled groups and other services, improve the level of service quality supervision and improve the transparency of services, safeguard social fairness through the information service terminal construction on the basis of improving coverage scale

(continued)

Table A.1 (continued)

First grade indicator	Secondary indicators	Tertiary indicators	Indicator description
		Medical services	Refers to improve the level of basic public health services through the application of information technology; guarantee all kinds crowd such as children, women, the old man get satisfactory service through the information management system construction and terminal services; guarantee the safety supply of food and drug through the establishment of traceability system of food and drugs, and promote public opinion surveillance, improve the transparency of service quality supervision
		Public culture and sports	Refers to enlarge the services of public cultural service area, improve the popularity rate of access radio, film and television through the application of information technology; improve the convenient degree of cultural content for all kinds of people, through popularity of information terminal; and increase the sports facilities service's coverage and utilization
		Services for disabled	Refers to promote the level of social security and basic services for the disabled, provide a sound body, health service facilities and rich service content through the Informatization, personalized application development on the basis of improving the coverage rate of service
Smart management and service	Special application	Basic housing security	Refers to through the application of information technology, the promote the service level of low-rent housing, public housing areas, shantytowns transformation, enhance the convenience of service, improve the transparency of the service
		Intelligent transportation	Refers to the smart construction and operation of city's whole traffic, including public transportation construction, traffic accident treatment, application of electronic maps, the urban road construction of sensor and the traffic induced information application
		Smart energy	Refers to the construction of urban energy wise management and use, including smart meter installation, energy management and utilization, the construction of street lamp intelligent management, etc.
		Smart environmental protection	Refers to the construction of urban environment, management and service of ecological wisdom, including the construction of air quality monitoring and service, surface water environment quality monitoring and service, the environmental noise monitoring and service, pollution sources monitoring, urban water environment, etc.

(continued)

Table A.1 (continued)

First grade indicator	Secondary indicators	Tertiary indicators	Indicator description
		Smart land	Refers to the smart construction of the city land and resources management and service, including the construction of land use planning implementation and monitoring of land resources, land use change monitoring, cadastral management, etc.
		Smart emergency	Refers to the construction of urban smart emergency, including the construction of emergency relief materials, emergency response mechanism, emergency response system, disaster warning capabilities, disaster prevention and mitigation capabilities, emergency command system, etc.
		Smart safety	Refers to the smart construction of urban public safety system, including urban food safety, drug safety, safe city construction
		Smart logistics	Refers to the construction of intelligent logistics management and service, including the construction of logistics public service platform, intelligent warehousing services, logistics call center, logistics traceability system, etc.
		Smart community	Refers to the digital, convenient, intelligent level of the community management and service, including the construction of community service information push, information service system coverage, community sensor installation, community operation security, etc.
		Smart home furnishing	Refers to the construction of home furnishing safety, convenience, comfort, artistic and environmental protection and energy conservation, including home furnishing intelligent control, such as intelligent appliance control, lighting control, security control and access control, home furnishing digital service content, home furnishing facilities installation, etc.
		Smart payment	Refers to smart new payment way of one cartoon, mobile payment, citizen card, and construction of convenience, safety, and construction of merchants pay for convenience, safety, etc.
		Smart finance	Refers to the urban financial system smart construction and services including the construction of honest regulatory system, investment and financing system, financial security system, etc.

(continued)

Table A.1 (continued)

First grade indicator	Secondary indicators	Tertiary indicators	Indicator description
Smart industry and economy	Industry planning	Industry planning	Refers to the urban industrial planning-making and completion, surrounding the city industry development, industry transformation and upgrade, the strategic industries of emerging industry development planning, planning of the public and the situation of the implementation
		Innovation input	Refers to the city's innovation industry input, including cost input of industry transformation and upgrade, innovation input of the development for emerging industry, etc.
	Industry upgrade	Industrial elements gathered	Refers to the city for the industry development, industry transformation and upgrade and the implementation of industry elements gathered, growth conditions
		Transformation of tradiitional industry	Refers to realize the transformation of traditional industries in the process of achieving industrial upgrading
	Development of emerging industry	High and new technology industry	Refers to the services and development of city's high and new technology industries, including the talents environment, scientific research environment, financial environment and the status of the management service industries, the development of high-tech industries and the status of the level of the whole industry in the city supporting the new and high technology
		Modern service industry	Refers to the city's modern service industry development status, including the policy environment for the, development environment, development level and investment of the development of modern service industry
		Other emerging industry	Reflecting the development and the status of the promotion of the city's other emerging industries

Table A.2 MIIT's intelligent city evaluation indicator system

Overall indicators	First grade indicator	Secondary indicators	Inspection points		Description
City intelligence	Intelligence preparation	Network environment	Fixed broadband	Average rate of the network	Comprehensively reflected the construction of fixed broad band's applied level
				User ratio of using 4M broadband products or above	
				Household rate of optical fiber	
				Internet penetration rate	
			Mobile internet	3G network coverage	Comprehensively reflected the construction of mobile internet's developmental level
				WLAN coverage	
				Owning rate of smart phone	
				User ration of Mobile broadband	
		Technical preparation	Internet of Things application demonstration	Internet of Things application demonstration efficacy in key industries of economic operation, including industry, agriculture, circulation industry	Comprehensively reflected the construction of Internet of Things in the city's key field
				Internet of things application demonstration efficacy towards the fields of infrastructure and security assurance, including transportation, electricity, environmental protection	
				Internet of things application demonstration efficacy towards the fields of social management and the people's livehood services, including public safety, health care, smart home	
City intelligence	Intelligence preparation	Technical preparation	Cloud computing technology application	Effectiveness of demonstration projects related to cloud computing applications	Reflecting the application of cloud computing technology
				Wether to formulate policy documents to guide cloud computing applications	
				Financial support for cloud computing applications	
		Guarantee conditions	Policy planning	Wether to formulate documents related to intelligent city's development outline, special plan, action plan	Reflecting the government attention to the construction of intelligent city

(continued)

Table A.2 (continued)

Overall indicators	First grade indicator	Secondary indicators	Inspection points		Description
				Wether to formulate relevant policy documents of encouraging the development of city informatization and application	
			Capital talents	Government investment, social investment and financing support related to the intelligent city construction	Reflecting the guarantee capability of intelligent city related capital talents
				Number of city information technology professionals, and the developing training number of high school and training agencies' related talents	
	Smart management	City operation management ability	Number of unit area information collection, monitor equipment		Reflecting city's integrated information collection ability
			Construction and application of the business support system or operative office, resource sharing, administrative examination and approval, administrative law enforcement supervision		Reflecting the system support ability of the government integrated service
			Construction and application of business support system for economic monitoring, credit supervision, investment and financing, energy saving and emission reduction		Reflecting the system support ability of city economy operating business
			Construction and application of business support system for traffic, social security, medical care, education, environmental protection		Reflecting the system support ability of city social business management
			Construction and application of business support system for water supply, power supply, gas supply, land resources		Reflecting the system support ability of city municipal resource management
			Construction and application of business support system for public security, emergency, civil air defense, dangerous goods management		Reflecting the system support ability of city public safety management
			Ability of collecting comprehensive operation management data, analysis and processing, support the urban management decision-making		Comprehensively reflect the system's leading decision-making ability

(continued)

Table A.2 (continued)

Overall indicators	First grade indicator	Secondary indicators	Inspection points	Description
			Ability to use electronic means in the progress of administrative law enforcement, administrative examination and approval, public service provision for supervision and inspection	Reflecting the ability of electronic supervision in the process of city management
			Application effect of using smart terminal devices to improve daily office efficiency, and carry out city management	Reflecting the mobile office ability in the process of city operating management departments
		Controlling of construction process	Whether or not to develop documents, standards to guide the construction management of intelligent city project	Reflecting the standardization and normalization of the process management of the intelligent city project construction
City intelligence	Smart management	Process control of construction	Schedule deviation rate = (actual time of completion − planned time of completion)/planned time of completion	Reflecting the progress deviation degree of major projects and major programs
			Budget deviation rate = (actual investment − budget amount)/budget amount	Reflecting the budget deviation degree of major projects and major programs
		Operation management mode	Whether the operation management subject is clear	Reflecting the perfect degree and maturity of the intelligent city operating management model to a certain extent
			Guarantee of the operating management capital	
			Whether formulating the system's standards related to the operating management	
	Smart services	Smart service coverage	Proportion of realizing online handling in real time in the city administrative service matters	Reflecting the coverage level of the online administrative service
			Proportion of realizing online real time provided in the city public service matters	Reflecting the coverage level of the online public service
			Whether disclosing the important information of financial capital usage, personnel, statistics of funds to the public in real time through the government website	Reflecting the coverage level of government website's important information

(continued)

Table A.2 (continued)

Overall indicators	First grade indicator	Secondary indicators	Inspection points	Description
			Whether the fields of medical treatment, social security, education, employment and traffic provide real time online consulting complaints in accordance with the user's needs	Reflecting the coverage level of major field's online consulting complaints
		Convenience of accessing	The development status of the government website and related public service websites	Accessing convenience of the city services
			Construction level of the government mobile application and public service mobile application	
			Whether can analyze user's needs, actively push the related information and service	
			Whether can surround customer service requirements, associating the related service items and information resources, to show to the users	
		Treatment efficiency	Average handling time of the approval hall service matter	Comprehensive reflects the working efficiency of the city
			Average reply time of business consulting	
			Average processing reply time of complaints	
			Response speed of online service system	

Table A.3 The first Guomai smart city developmental level evaluation

First grade indicator	Secondary indicators	Tertiary indicators
Smart infrastructure	Information network facility	Broadband network
		Integration of three networks
	Information sharing infrastructure	Public cloud computing center
		Information safety service
		Government affairs cloud
Smart infrastructure	City infrastructure	Intelligent transformation in key areas
Smart governance	Smart government affairs	Decision-making ability
		Government services and transparency
		Business collaboration level
	Smart public management	Smart transportation
		Smart city inspectors
		Smart pipe network
		Smart security and protection
		Smart food and drug administration
		Public and social participation
Smart people's livehood	Smart social insurance	Social security system construction level
		Social security information service level
	Smart health security	Health security information service level
	Smart education culture	Education culture information service level
	Smart community services	Community information service level
Smart industry	Per-capita output	Per-capita output
	Input-output ratio	Input-output ratio
	Resource consumption rate per ten thousand GDP	Resource consumption rate per ten thousand GDP
	Integration of the two	Environment of the integration of the two
		Level of the integration of the two
		Benefit of the integration of the two
Smart crowd	Information utilization capacity	Application of the information products
		Utilization of the information resources
	Innovation ability	Innovation environment
		Knowledge innovation ability
	Quality of talents	Higher education situation

(continued)

Table A.3 (continued)

First grade indicator	Secondary indicators	Tertiary indicators
Smart crowd	Quality of talents	Senior talent status
		Talent introduction status
Smart environment	Ecological protection	Environmental construction level
		Environmental informatization level
	Resource utilization	Resource saving level
		Intelligent resource application
	Soft environment construction	Organizational system
		Planning policies
		Regulatory standards
		Evaluation and assessment
		City brand

Table A.4 2nd smart city development level assessment of Guomai Company

First grade indicator	Secondary indicators	Evaluation points	Evaluation description
Smart infrastructure	Coverage of optical fiber and broadband	Household rate of optical fiber	Household ratio of family optical fiber
		Wi-Fi coverage	Coverage of all kinds of wireless transmission networks in city areas
	Popularizing rate of computer terminal	Netizen number	Number of Internet users in the proportion of urban population
	Cloud platform	Construction or utilization status of cloud computing center	Wether to plan or have been established (rented) cloud computing center
Smart application	Typical applied project	Citizen card	Issuing proportion of citizen card
		Filing rate of residents' health records	Filing rate of residents' health records
		Smart electric meter mounting yield	Proportion of installing smart electric meter in households
		Other typical application	Other typical under construction or demonstrated application projects (such as achievement, tourism, safety, community and environment)
Smart industry	Industrial development level	Per-capita output value	Gross national product per capita
		Electric power consumption per capita	Citizen electric power consumption per capita
		Number of per capital patent (million)	Patent authorization quantity of million citizens
		Proportion of high and new technology industry output value to GDP	Proportion of high and new technology industry output value to GNP

(continued)

Table A.4 (continued)

First grade indicator	Secondary indicators	Evaluation points	Evaluation description
Smart governance	Government service ability	Public service platform	Whether there's a special platform with public web pages, convenience services, service hall or enterprise oriented
		Integrity of government information disclosure	Timeliness and effectiveness of government information on the official website
		Online service ability	Universality and convenience of government online service
		Information resources utilization	Government portal website daily visits per capita
Intelligent support ability	Planning scheme	Overall plan or action program	Established intelligent city detailed planning outline or action
			Upcoming intelligent city planning or outline that formulated in progress
			The formed f overall intelligent city development of ideas, such as wireless city, city optical network
	Organization system	Special leading institutions or actuators	Mayor/secretary of the leadership positions
			Position above city deputy, and lead as a leader or executive body
			Position bellow city deputy, and lead as a leader or executive body and other cases
	Investment	Special fund budget	Issued and clear intelligent city special fund budget
			Investment in the construction of intelligent city special funds by operators or integrators
	Media and promotion	A variety of promotional activities	Held an intelligent city related training, seminars, conferences or forums, such as the creation of topics, websites and micro-blog

Table A.5 4th Guomai intelligent city development level evaluation

First grade indicator (6)	Weight	Secondary indicators (15)	Weight
Smart infrastructure	25	Broadband	10
		Basic database completeness	5
		Application of city cloud platform	10
Smart management	20	Government collaboration level	5
		Implementation of industry's total solution	10
		Public management social participation	5
Smart service	20	Integrated people's livelihood service capacity	10
		Government data open service	10
Smart economy	15	Number of patent per capita	5
		Energy consumption per ten thousand GDP	5
		Proportion of information industry value added to GDP	5
Smart crowd	10	Proportion of 3G to 4G users	5
		Consumption of e-commerce per capita	5
Security system	10	Formulation status of development planning	5
		Organization and performance evaluation	5
Plus (1)	5	Intelligent city's pilot construction and application innovation, related honors and major events	5
Total	105		105

Table A.6 Engineering research institute intelligent city (town) development index

First grade indicator	Secondary indicators
Happiness index of intelligent city	Employment income
	Cultural education
	Medical services and health
	Social insurance
	Housing and consumption
	City cohesion
	Public service
	Organization and infrastructure
	Social service
Smart city management index	Economic base
	Science and technology innovation level
	Manpower resource
	Human settlement
	Environmental action
	Ecological environment
Intelligent city social responsibility index	Level to govern
	Region influence
	Image transmissibility
	Management and decision
	Public responsibility
	Equity responsibility
	Integrity responsibility

Table A.7 Pudong smart city indicator system 1.0

First grade indicator	Secondary indicators	Tertiary indicators	Reference value
Infrastructure	Coverage level of broadband network	Access rate of family fiber	≥ 99%
		Coverage rate of wireless network	≥ 95%
		Wlan coverage rate of main public	≥ 99%
		Coverage rate of next generation broadcast	≥ 99%
	Broadband network access level	Network access level per household	≥ 30
		Average wireless network access bandwidth	≥ 5M
	Infrastructure investment level	Proportion of basic network facilities investment in fixed assets investment	≥ 5%
		Construction level of sensor network (total investment in fixed assets)	≥ 1%
Public management and service	Smart government service	Administrative examination and approval project online management proportion	≥ 90%
		Electronic monitoring rate of government official behavior	100%
		Rate of net transfer of non-official documents	100%
		Network interaction rate between corporate and government	≥ 80%
		Network interaction rate between citizens and government	≥ 60%
Public management and service	Smart transportation management	Concern rate of citizen to the traffic information	≥ 50%
		Electronic rate of bus stop board	≥ 80%
		Citizen traffic guidance information compliance rate	≥ 50%
		Coverage rate of parking guidance system	≥ 80%
		City road sensing terminal installation rate	100%
	Smart medical treatment system	Filing rate of citizen electronic health record	100%
		Electric medical history usage rate	100%
		Resource and information sharing rate between hospitals	100%

(continued)

Table A.7 (continued)

First grade indicator	Secondary indicators	Tertiary indicators	Reference value
	Smart environmental protection network	Monitoring proportion of environment quality automation	≥ 95%
		Monitoring proportion of key pollution source	100%
		Discharge index of carbon (declining compared with 2005)	≥ 40%
	Smart energy management	Family smart meter installation rates	≥ 50%
		Enterprise intelligent energy management proportion	≥ 70%
		Intelligent management proportion of road lamp	≥ 90%
		New energy car's proportion	≥ 10%
		Building digital energy saving ratio	≥ 30%
	Smart city security	Food and drug traceability system coverage rate	≥ 90%
		Natural disaster early warning release rate	≥ 90%
		Construction rate of major emergency response system	100%
		Coverage rate of city grid management	≥ 99%
		Household population and permanent population information tracking	≥ 99%
	Smart educational system	Urban per capita expenditure on education (GDP)	≥ 4.5%
		Information interaction rate of family and school	≥ 90%
		Online teaching proportion	≥ 50%
	Smart community management	Coverage rate of community information service system	≥ 99%
		Published rate of community service information	≥ 95%
		Coverage rate of information service for community elderly	≥ 90%
		Residential area safety monitoring sensor installation rate	≥ 95%
Information service economy development	Industry development level	Value added of information services accounted for GDP	≥ 10%
		Proportion of e-commerce transactions to total sales of commodities	≥ 30%

(continued)

Table A.7 (continued)

First grade indicator	Secondary indicators	Tertiary indicators	Reference value
		Proportion of the information service industry employees in total social employees	≥ 10%
	Enterprise informatization operation level	Fusion index of industrialization and informatization	≥ 85
		Enterprise website establishment rate	≥ 90%
		Enterprise e-commerce behavior rate	≥ 95%
		Utilization rate of enterprise information system	≥ 90%
Quality of humanistic social science	Citizen income level	Per capita disposable income (RMB)	≥ 50,000 Yuan
	Citizen culture science literacy	Proportion of college or above in total population	≥ 30%
		Standard-reaching rate of city public scientific literacy	≥ 20%
	Public information publicity and training level	Proportion of the relevant publicity and training of the total population every year	≥ 8%
	Citizen's life networking level	Rate of the citizen's surf the internet	≥ 60%
		Usage proportion of mobile internet	≥ 70%
		Family online shopping proportion	≥ 60%
Citizen's subjective perception	Easy feeling of life	Satisfaction of network charge	≥ 8 points
		Convenience of accessing traffic information	≥ 8 points
		Convenience of city's seeking for medical service	≥ 8 points
		Convenience of government service	≥ 8 points
		Convenience degree of accessing education resource	≥ 8 points
	Sense of safety for life	Food and drug safety satisfaction	≥ 8 points
		Environment safety satisfaction	≥ 8 points
		Traffic safety satisfaction	≥ 8 points
		Prevention and control of crime satisfaction	≥ 8 points

Table A.8 Pudong smart city evaluation indicator system 2.0

First grade indicator	Secondary indicators	Tertiary indicators
Infrastructure	Broadband network's construction level	Access rate of family fiber
		WLAN coverage rate of main public areas
		Network access level per household
Public management and service	Smart government service	Administrative examination and approval level online
		Rate of net transfer of non-official documents
	Smart transportation management	Electronic rate of bus stop board
		Citizen traffic guidance information compliance rate
	Smart medical treatment system	Filing rate of citizen electronic health record
		Electric medical history usage rate
	Smart environmental protection network	Monitoring proportion of environment quality automation
		Monitoring proportion of key pollution source
	Smart energy management	Family smart meter installation rates
		Proportion of new energy automobile
		Buildings digital energy saving ratio
	Smart city safety	Construction rate of major emergency response system
		Hazardous chemicals transportation monitoring level
	Smart education system	City education spending level
		Online education proportion
	Smart community management	Community integrated information service ability
Information service economy development	Industry development level	Value added of information services accounted for GDP
		Proportion of the information service industry employees in total social employees
	Enterprise informatization operation level	Enterprise website establishment rate
		Enterprise e-commerce behavior rate
		Utilization rate of enterprise information system

(continued)

Table A.8 (continued)

First grade indicator	Secondary indicators	Tertiary indicators
Quality of humanistic social science	Citizen income level	Per capita disposable income
	Citizen culture science literacy	Proportion of college or above in total population
Quality of humanistic social science	Citizen's life networking level	Rate of the citizen's surf the internet
		Family online shopping proportion
Citizen's subjective perception	Easy feeling of life	Convenience of accessing traffic information
		Convenience of city's seeking for medical service
		Convenience of government service
	Sense of safety for life	Food and drug safety electronic monitoring satisfaction
		Environmental safety information monitoring satisfaction
		Satisfaction of traffic safety information system
Intelligent city soft environment construction	Smart city plan and design	Intelligent city develops and plan
		Intelligent city leadership system
	Intelligent city atmosphere building	Intelligent city forum meeting and training level

Table A.9 Smart Nanjing evaluation indicator system

First grade indicator	Secondary indicators
Fundamental areas	Coverage rate of wireless network
	Fiber access coverage
	Average network bandwidth
	Number of national key laboratories
	Smart grid technology and equipment applications
Smart industry	Smart industry fixed assets investment
	Smart industry's R&D appropriation expenditure
	Proportion of smart industry to GDP
	Number of smart industry practitioners
	Total application number of patent in smart industry
	E-commerce transactions
	GDP energy consumption of million Yuan
Smart service	Government administrative efficiency index
	Collaborative application system
	Smart public service widespread application
	Smart service construction funds
Smart humanity	City labor productivity
	Proportion of junior college degree or above
	Proportion of information service industry practitioners to the whole social practitioners
	Total information level indexes
	City public service satisfaction survey
	Proportion of cultural creative industry to GDP
	Evaluation on international cultural and sports activities

Table A.10 Ningbo intelligent city development evaluation indicator system

First grade indicator	Secondary indicators	Tertiary indicators
Smart crowd	Manpower resource	Number of higher education every ten thousand people
		Number of technical personnel every ten thousand people
		Proportion of information industry practitioners to the whole social practitioners
	Lifelong learning	Public library books and documents checked out per capita
	Information consumption	Information consumption coefficient per capita
		E-commerce transactions per capita
Smart infrastructure	Communication facilities	Mobile phone holding number every one hundred people
		Cable TV two-way digital transformation rate
		Computer holding quantity every hundred households
		Cable broadband access rate
	Information sharing infrastructure	Wireless broadband network coverage
		Construction status of government data center, four basic databases, information security disaster
Smart infrastructure	Information sharing infrastructure	Communication network sharing and co-building
		Digital management level of infrastructures
Smart governance	E-government affairs	Government affairs weibo number
		Status of one-stop online administrative approval service and electronic monitoring system construction
		Hits number of city government portal website
	Public participation in government decision-making	Number of NPC bill registered
		Number of CPPCC proposal registered
		Number of public hearing
	Input of public service	General public service expenditure (local finance)
Smart people's livehood	Social security	Status of social security and health insurance one-card construction
		Status of citizen card project construction

(continued)

Table A.10 (continued)

First grade indicator	Secondary indicators	Tertiary indicators
	Medical treatment	Online booking hospital proportion
		Filing rate of resident's electronic health record
	Transportation	Transportation card per capita
		City traffic guidance system
		Electronic rate of bus stop board
Smart economy	Economy power	Regional total output value per capita
	Smart industry	Proportion of information industry added value to GDP
		Proportion of software outsourcing services to GDP
	R&B ability	Weight of R&B to GDP
		Patent authorization quantity of million persons
	Output energy consumption	Energy consumption of GDP per million Yuan
	Industrial structure and contribution	Average added value of the agriculture, forestry, animal husbandry and fishery created by the employee
		Proportion of high technology added value above the scale to the industrial added value
		Proportion of the added value of the third industry to GDP
Smart environment	Disposal capability	Town life sewage treatment rate
		Comprehensive utilization rate of industrial solid wastes
	Environment attraction	Comprehensive utilization of "three wastes" product output
		Greening coverage of built areas
		Per capita green area
Smart planning and construction	Integration of urban and rural overall development	Residents income ratio between city and country
		Education years proportion of urban and rural residents
		Public finance spending proportion of urban and rural
		Urbanization rate
	Spatial arrangement	Commuting time (or transfer number)
	Smart buildings	Building intelligent level

Table A.11 TU Wien indicator system

Dimension	Factor (31 items)	Weight (%)	Indicator
Smart economy	Innovative spirit	17	Weight of R&D to GDP
			Employment rate of knowledge intensive industry
			Patent application per capita
	Entrepreneur spirit	17	Weight of professional
	Entrepreneur ship		New enterprise registration number
	Economy prospect	17	Importance as the decision center
	Economic image and trademarks		
	Productivity	17	City labor productivity of employed population
	Flexibility of labour market	17	Unemployment rate
			Part-time employment rate
	Degree of internationalization	17	Total number of listed companies
	International Embeddedness		Air passenger flow volume
			Air freight traffic volume
Smart citizen	Level of qualification	14	Knowledge center importance (top research centers and universities)
Smart people			Population of level ISCED5 ~6 (above colleges and universities)
			Level of foreign language
	Participation of life long learning	14	Number of borrowing books per capital
			Participation rate of lifelong learning
			Participation rate of language course
	Social race diversity	14	Foreigner weight
		14	Proportion of citizens born overseas
	Flexibility	14	Attitude to job-hooping
	Creativity	14	Creative crowd weight
	Openness	14	Status of participating in European election
			Immigrant environmental friendly degree
			Understanding of the Europe

(continued)

Table A.11 (continued)

Dimension	Factor (31 items)	Weight (%)	Indicator
	Participation in public life	14	Status of Participating in city election
			Status of Participating in volunteer work
Smart governance	Participation in decision-making	33	Number of citizen representative per capita
			Residents' political activities
			Importance of political for residents
			Proportion of women representatives
	Public and social services	33	Public expenditure per capita
			Day care children proportion
			School quality satisfaction
	Transparent governance	33	Satisfaction of government transparency
			Anti-corruption satisfaction
Smart transportation	Local accessibility	25	Public transport network per capita
			Accessibility satisfaction of public traffic
			Public transport quality satisfaction
	(Inter-)national accessibility	25	(Inter-)national accessibility
	Availability of ICT-infrastructure	25	Average computer for each household
			Average broadband for each household
	Sustainable, innovative and safe transport systems	25	Green traffic weight
			Traffic safety
			Economical car use
Smart environment	Attractively of natural environment	25	Sunlight duration
			Green space weight
	Pollution	25	Thermal aerosol (ozone)
			PM (particulate matter)
			Deadly chronic lower respiratory diseases

(continued)

Table A.11 (continued)

Dimension	Factor (31 items)	Weight (%)	Indicator
	Environmental protection	25	Individual environmental protection efforts
			Environmental protection attitude
	Sustainable resource management	25	Water consumptions per GDP
			Energy consumptions per GDP
Smart living	Cultural facilities	14	Per capita number to the cinema
			Per capita number to visit the museum
			Number to the theater per capita
	Health conditions	14	Expected lifetime
			Hospital beds per capita
			Doctor number per capita
			Medical system satisfaction
Smart living	Individual safety	14	Crime rate
			Mortality rate of violent crime
			Safety satisfaction
	Housing quality	14	Weight of meeting the lowest standard dwelling
			Per capita living space
			Personal housing conditions satisfaction
	Education facilities	14	Per capita number of students
			Educational facilities acquisitiveness satisfaction
			Education facilities quality satisfaction
	Touristic attractively	14	Importance to be the sightseeing
			Annual per capita number of overnight visitors
	Social cohesion	14	Poverty risk perception
			Poverty rate

Table A.12 IDC Indicator system

Dimension	Unit	Evaluation criterion
Smart dimension	Smart government	Open government
		Government/E-services
		Electronic service supply
		Sustainable behaviors
		Environmental protection policy
	Smart buildings	Buildings operating efficiency
		Construction quality
	Smart transportation	Power-driven transportation
		Traffic information and management
		City public transportation
Smart dimension	Smart energy and environment	Smart power grid
		Renewable energy
		Environment management
	Smart service	Public service/emergency services
		Tourisms/building/modern services
Supporting capacity	Information and communication technologies	Transmissibility
		Mobility
	Citizen	Age
		Education
		Population development
	Economy	Economic wealth
		Economic development

Table A.13 IBM indicator system

System	Element	Internet of Things	Interconnection	Intelligent
City service	Public service management/local government management	Creation of local authority management information system	Interconnected service delivery	Immediate and joint service provision
Citizen	Health and education public safety government services	Patient diagnosis and screening equipment	Connect doctors, hospitals and other health service providers' records	Patient driven early treatment

(continued)

Table A.13 (continued)

System	Element	Internet of Things	Interconnection	Intelligent
Commerce	Business environment management burden	Data collection about the online business services	Various stakeholders connecting city commercial system	Provide customized service for business
Traffic	Cars and road public transport airports and seaports	Use of measuring traffic flow and tolls	Integrated traffic, weather and traveling information services	Highway charges
Communication	Broadband, wireless telephone, computer	Collecting data by phone	Connect mobile phones, fixed telephone and broadband	Provide consumers with personalized city service information
Water supply	Health clean water supply salt water	Collecting water quality monitoring data	Connect the water supply enterprise, port, energy users	Quality, flood drought response
Energy	Oil and gas renewable energy nuclear energy	Using sensors to collect usage data in an energy system	Device and equipment connecting the energy consumer and supplier	Optimize the use of the system, and balance the usage of different time

Table A.14 Ericsson indicator system

Category	Field	First grade indicator	Secondary indicators
ICT development maturity	Infrastructure	Broadband quality	Average download speed
			Cell-edge network quality
			Network band
		Accessibility	Family internet penetration
			Access rate of fiber
			High speed wireless network
			Number of Wi-Fi hotspots
	Acceptability	Charge rate	Proportion of broadband tariff to city labor productivity
			Proportion of mobile tariff to city labor productivity
		IP switching charge	IP switching charge per megabyte data
	Application	Science and technology application	Number of mobile phone
			Number of smart phones per capita
			Family computer penetration rate
			Number of tablet per capita
		Personal application	Per capita internet penetration
			Social network penetration
		Public and market application	Open data
			Electronics and mobile payment
Three bottom line effect	Society	Health	Children dead under 1 year old
			Average life
		Education	High school or college education degree
			Rate of education
		Inclusive	Murder in every 100 thousand residents
			Unemployment rate
			Gender equality of education

(continued)

Table A.14 (continued)

Category	Field	First grade indicator	Secondary indicators
	Economy	Efficiency	Gender equality in city council
			City labour productivity
		Competitiveness	Higher education popularity
			PCT patent of every million residents
			Knowledge-intensive employment rate
			New enterprises of every 100,000 residents
	Environment	Resource	Per capita garbage
			Recycling garbage per capita
			Per capita fossil fuel consumption
			Per capita non fossil fuel consumption
		Pollution	PM10 concentration
			PM2.5 concentration
			Nitrogen dioxide solubility
			Sulfur dioxide solubility
			Sewage treating rate
		Climate change	Carbon emissions per capita

A.2 Fuzzy Delphi Expert Consultation Questionnaire for Intelligent City Evaluation Indicator

During March 2013 to August 2013, the research group issued 56 questionnaires to the academicians and experts within the research group about the project "Strategic Research on Construction and Promotion of China's Intelligent Cities" of China Academy of Engineering, to revise the indicator and determine the indicator selected by expert scoring. Contents of the specific expert consultation questionnaire are as follows.

To _____:

It is a great honor to invite you to grant instruction on this questionnaire.

"Intelligent City Evaluation Indicator System Research" is a research topic that attaches to Chinese Academy of Engineering's key consulting project "Strategic Research on Construction and Promotion of China's Intelligent Cities", the evaluation object of this research is the city's intelligent construction and sustainable development, starting from five dimensions: smart environment and construction, smart management and service, smart economy and industry, smart hardware facilities, residents' intelligent literacy, and reflecting them by setting the corresponding secondary indicators and tertiary indicators according to a certain principle, respectively. Wherein, the set of the secondary indicators is mainly for the government agencies' working field, to reflect the cohesion of management services; the set of the tertiary indicators takes into account the data that can be obtained, and the development status and trend of development, to reflect the focus.

This questionnaire is an expert consultation questionnaire, which applied the fuzzy Delphi method, focusing on the importance of the secondary indicators and the tertiary indicators under the five dimensions of the Intelligent City Evaluation Indicator System.

You can choose the familiar parts of the questionnaire to fill, after you finish the chosen parts 1–5 in this questionnaire, please send the results back to us before July 25, 2013. Thank you again for your advice! If you have any further questions or suggestions, please contact us at any time.

"Intelligent City Evaluation Indicator System Research" Research Group

- Basic information

  Your major:_____

  Your title: ☐Academician ☐Professor ☐Associate professor ☐Other____

  Your age: ☐<30 years old ☐30–39 years old ☐40–49 years old ☐50–59 years old ☐ ≥ 60 years old

This questionnaire is an open questionnaire. Its purpose is to consult on the concern about the evaluation of intelligent city, thus to further evaluate the secondary and tertiary indicators in all dimensions, and to form the evaluation indicator can be evaluated. The evaluation of each concern contains the following two parts.

(1) Indicator concern points to fill: in addition to the secondary and tertiary indicators that have been filled out, you have to write down the indicators that should be concerned in the intelligent city evaluation in the two columns for each dimension, respectively.

(2) Importance assessment: for the existing or new indicators that you feel needed to be add, please evaluate the importance of the Indicators in the rear column, and tick the box in the degree of importance ($\sqrt{}$).

The above degree of importance has five grades: 1–5 points (the higher the score is, the more important it is), please evaluate the importance of each focus, and fill in the important level of the indicator.

Scores	1	2	3	4	5
Importance degree	A lot less important	Less important	General	Important	Very important

Example: Evaluation of ecological environment quality secondary indicators and tertiary indicators.

Secondary indicators	Tertiary indicators	Indicator increase or decrease/amending opinions	Importance Assessment				
			1	2	3	4	5
Ecological environment quality	Energy efficiency air pollutant concentration environmental pollution indicator	Per capita green area (added, importance 4) Water pollution index (added, importance 4)				$\sqrt{}$	
							$\sqrt{}$
				$\sqrt{}$			
					$\sqrt{}$		

- Questionnaire filling and explanation of the indicators

Consultation and evaluation object: intelligent city evaluation indicators.

Consultation and evaluation purpose: From the points of five first grade indicators (dimensions): smart environment and construction, smart management and service, smart economy and industry, smart hardware facilities, residents' intelligent literacy, to select and sort out the evaluation indicators of Intelligent City, and to establish a feasible, concise and sustainable evaluation indicator system.

- Filling the questionnaire

Please tick ($\sqrt{}$) to choose 1–5 familiar field(s) to fill out and evaluate the indicator's dimension

1. Indicators of the smart environment and construction dimension and its importance assessment

Secondary indicators	Tertiary indicators	Indicator increase or decrease/amending opinions	Importance Assessment 1 2 3 4 5
Resource performance	Energy consumptions per GDP Energy consumption per capita		
	Per capita water consumption Renewable energy ratio City labor productivity		
Ecological environment quality			
	Pollution monitoring coverage Air pollutant concentration Industrial pollutants of unit GDP		
City pollutants treatment			
	Reuse rate of sewage treatment Reuse rate of industrial waste treatment Reuse rate of domestic rubbish disposal Proportion of green transportation		
Built environment			
	Per capita public green areas City population density		

2. Indicators of smart management and service dimension and its important assessment

Secondary indicators	Tertiary indicators	Indicator increase or decrease/amending opinions	Importance Assessment 1 2 3 4 5
Residents demand guarantee	Coverage rate of supplying water Per capita living space		
Medical and health service	Electric medical history usage rate Expected life		
Degree of the security of city			
Public participation proportion	Crime rate		
	Percentage of voter turnout		
	Public participation proportion		
Public service satisfaction	Public satisfaction with the government		

3. Indicators of smart economy and industry dimension and its important assessment

Secondary indicators	Tertiary indicators	Indicator increase or decrease/amending opinions	Importance Assessment 1 2 3 4 5
Input and output efficiency			
Industrial development trend	Investment in GDP ratio City output density		
	Proportion of GDP three industries Urban labor productivity		
Ratio of capital investment			
Information service industry	External investment ratio Proportion of R&D spending to GDP		
	Proportion of information service industry professionals		
	Proportion of information service GDP		
Industry contribution per capita			
	City labor productivity		

4. Indicators of smart hardware facility dimension and its important assessment

Secondary indicators	Tertiary indicators	Indicator increase or decrease/amending opinions	Importance assessment 1 2 3 4 5
Information technology infrastructure			
	Information technology infrastructure investment proportion Access rate of family high speed network WLAN coverage rate in public space		
Construction of informatization of human resources			
	Proportion of information service industry professionals		
Information technology application			
	Proportion of mobile board band Proportion of Internet Per capita network information searched volume		

5. Indicators of residents' intelligent literacy dimension and its important assessment

Secondary indicators	Tertiary indicators	Addition or deletion of indicators / Amending Opinions	Importance Assessment				
			1	2	3	4	5
Social spending							
	Proportion of education spending to budget						
Residents' education degree							
	Education rate of senior high school Proportion of junior college degree or above Lifelong learning engagement						
Development of social justice	Gini coefficient						
Society diversity							
	Proportion of immigration						
City innovation							
	Proportion of creative industries to GDP Number of universities and research institutes						

Other indicators setting suggestions:

_____
_____
_____
_____
_____
_____
_____
_____
_____
_____
_____
_____
_____
_____
_____

Thank you for your valuable suggestions and comments!

A.3   Instruction on Intelligent City Evaluation Indicator System

A.3.1 Intelligent Environment and Construction

- Indicator F11: City PM2.5/PM10 monitoring stations density

This indicator refers to the distribution density of PM2.5 and PM10 monitoring stations in the city, reflecting the level of city's perception to the environmental quality.

The issue of city's air quality has appeared in the process of rapid urbanization, such as London, the world's earliest industrialized country, suffering a serious air pollution in the 1950s, with the annual average of "fog day" (days when the visibility is not more than 1000 m) up to 50 days or so. During December 5–10, 1952, "London Smog Incident" occurred. The opera "La Traviata" was suspended for the audience could not see the stage, people in the theater were forced to leave, the dense smoke cut visibility to inches in the day time, and the land and water traffic almost paralyzed.[1] The United States also encountered similar problems between 1940 and 1970, Los Angeles Photochemical Smog Episode caused more than 400 deaths of people more than 65 years old, and many people developed symptoms of eye pain, headache, and dyspnea.[2]

Although there are many developed countries' warning taken from the overturned cart in front, the city air quality crisis still broke out at the end of 2012 in China. "Haze" has become a keyword of the year. In January 2013, 4 large-scale haze processes shrouded 30 provinces (autonomous regions and municipalities); Beijing only has 5 non-haze days. On January 4, 2014, National disaster reduction office and Ministry of Civil Affairs brought the haze harmful to the health into the natural disaster in 2013 and reported it.[3]

City air quality is related to people's daily living environment and physical and mental health, which is an important aspect of city intelligent construction while the density of urban PM2.5/PM10 monitoring points reflects the city's perception of environmental quality. Although at the present stage, there's no completely correlation between PM2.5/PM10 monitoring points density and city air quality, the city of high PM2.5/PM10 monitoring points density city will be more convenient and accurate to sense the quality of the environment, and take it as a basis for reaction of regulation.

This is conforming to the characteristics that the intelligent cities should be perceived, judged and be reactive.

Air quality query tool "National Air Quality Index" board casts air quality index of more than 400 major cities throughout the country in real time, with the data

---

[1]For details see http://en.wikipedia.org/wiki/Great_Smog.
[2]For details see http://amuseum.cdstm.cn/AMuseum/atmosphere/main/k491.html.
[3]For details see http://www.gov.cn/zhengce/2014-02/13/content_2603649.htm

coming from Chinese Ministry of Environmental Protection and the U.S. Embassy[4]; the real time air quality index map provides the worldwide air quality indexes.[5] This book obtained the number of domestic and international city air monitoring points with the above method, respectively. To eliminate the influence of the different size of cities and the different definitions of size of cities at home and abroad, the size of urban built-up area is chosen as the standard of density measurement in the intelligent city evaluation indicator system. The data of the urban built-up area are obtained from the China Urban Statistical Yearbook, while for the foreign cities the sizes of urban area indicated on each city's official website were adopted.

City PM2.5/PM10 monitoring point density is obtained by the number of PM2.5/PM10 monitoring points within the city built-up area, and the unit is /km^2.

- Indicator F12: Level of city grid management coverage

The indicator refers to the proportion of management area of the unit grid divided in by the city management area in accordance with a certain standard in the total city management area, reflecting the level of city digital management.

In 2004, the pilot of domestic city grid was first created in Beijing, Dongcheng District, and in 2005, 51 cities promoted the pilot throughout the country. As of June 2012, among the 667 prefecture level and above (including prefecture level) cities throughout the country, there were 285 cities in the progress of city grid management. As a digital city management model, the feature of city grid is comprehensive utilization of mobile communication and network map and other high-tech means, to carry out all-round and efficient city management activities on the basis of control theory.

The main method of city grid management is to establish the grid electronic map corresponding to the city solid space, divide the city area fine grids on it, and designate them into several control areas according to a certain management range. Public components and events in the area are in accordance with their geographical location coded on the electronic map (Yan 2006).

Smart management of foreign cities usually combines the city grid electronic map with the city's open data, to carry out fine, dynamic management to the cities based on digital technology.

Since the grid management coverage is difficult to count, this book took the gradient evaluation method in calculation, with 100 (excellent level) indicating having been implemented, with 50 (general level), indicating that there is no specific implementation but broadcasted, part regions of part fields with implementation, and with 0 indicating having not been implemented. Construction management level of the intelligent city is evaluated by obtaining the information of city grid management platform, city geographic information management platform, city open data platform.

---

[4]For details see: http://air.fresh-ideas.cc.
[5]For details see: http://aqicn.org/map.

- Indicator F13: Residents' intelligent transportation tools usage level

The indicator refers to the degree of using intelligent traffic system and its auxiliary system when traveling, such as bus query system, real-time traffic system.

With the increase of urban traffic, the environmental and social problems caused by it are becoming more and more obvious. Intelligent transportation solutions are aimed at the use of a variety of technologies, to guide a reasonable traffic order, to ease urban traffic pressure, to reduce environmental pollution caused by motor vehicles. Meanwhile, it's also a very important part of people's lives. Good intelligent transportation system allows people to travel more fast, thereby improving the efficiency of work and life, directly makes people feel the convenience of the intelligent city.

Smart transportation system evaluated in this book shall include all aspects of before travel, traveling and arrival, involving Electronic Traffic Controlling System, Parking Guidance and Information System, Intelligent Transportation Cloud Information Service Platform, Smart Car Share Service System, Smart Road Surveillance and Maintenance System (SRSMS), Smart Bicycle/Vehicle and etc.

This book uses a gradient evaluation method in calculation, with 100 (excellent level) indicating having been implemented, with 50 (general level), indicating that there is no specific implementation but broadcasted, part regions of part fields with implementation, and with 0 indicating having not been implemented.

- Indicator F14: Level of the publicity of the city's online construction plan

This indicator refers to the publicity degree of intelligent city construction plan on government website, reflecting the public degree of intelligent city construction to the citizens.

With the development and popularization of Internet and mobile communication technology, more and more people have access to news and information through government public website and official accounts. Government website has become an important window for the government to promote new policies and measures, which needs to be built, maintained and updated in time.

As of the end of December 2013, the London City Council released about 50 pages of "Smart London Plan", and created "Smart London Vision" volume on the government webpages, fully publicize the concept of smart construction from data acquisition, display technology, the public impact. Compared to foreign cities' degree of openness for future plan on the Internet, there are few cities in China to publish detailed online intelligent city construction program.

This book uses a gradient evaluation method in the calculation, with 100 (excellent level) indicating having been published in detail, with 50 (general level), indicating that that there is no specific content but published, and with 0 indicating having not been published.

A3.2 Intelligent Smart Management and Services

- Indicator F21: Online publicity level of the government's non-classified documents

The indicator refers to the proportion of the number of government non-secret documents published on the website in the total documents, reflecting the transparency of government information.

As the basis of building an intelligent city, urban information acquisition and sharing is an important support for public decision-making. For government management, it should open up the various departments, the enterprise information access and sharing channels. Communications, transportation, health care, education, real estate and other public information are important resources for the construction of intelligent city; the government should provide an open, free, transparent information sharing. The profound changes in information and communication technology have changed all the social relations, including the relationship between the public and the government, and reshaped the way of public governance.

In 2013, Tsinghua University released the "2013 Chinese municipal government fiscal transparency report", which comprehensively evaluated the 289 cities (including 4 municipalities and 285 prefecture level cities), and established a full-sized indicator system for the evaluation of these cities' government fiscal transparency.[6] In this book, the government non-confidential documents' transparency of the domestic cities will be applied the evaluation content of this report on.

Transparency International published Corruption Perceptions Index every year.[7] The degree of transparency of the government is closely related to whether or not it implements the accountability system, to ensure the integrity of public services (Table A.15).

Continuous monitoring and review of government operations and plans, and deliberate eliminating the fragmented situation among departments, is the basis of openness, transparency and information sharing. This book adopts the data in the statement to evaluate the foreign cities, and replaces the evaluation of the cities with ones of the countries in which the cities are.

- Indicator F22: Online public participation ratio

The indicator refers to the proportion of public participation in the decision-making of city construction related events, reflecting the public participation in city construction and the openness, fairness and inclusiveness of decision-making.

Public participation is a kind of public participation in city construction decision-making activities, is a two-way exchange of views, which can widely

---

[6]For details see: http://www.sppm.tsinghua.edu.cn/eWebEditor/UploadFile/20130812031437755.pdf.
[7]For details see: http://issuu.com/silch/docs/2013_cpibrochure_en.

Table A.15 Top 30 scores of municipal government in 2013 China municipal government fiscal transparency report

Ranking	City	Total scores	Ranking	City	Total scores
1	Shanghai	45	16	Yulin (Guangxi)	35
2	Beijing	43	17	Anqing (Anhui)	34
3	Guangzhou (Guangdong)	43	18	Jincheng (Shanxi)	34
4	Changzhi (Shanxi/0	43	19	Haikou (Hainan)	33
5	Erdos (Inner Mongolia)	42	20	Jieyang (Guangdong)	33
6	Zhuhai (Guangdong)	41	21	Yueyang (Hunan)	32
7	Shenzhne (Guangdong)	41	22	Foshan (Guangdong)	32
8	Sichuan (Chengtu)	40	23	Nanning (Guangxi)	32
9	Hangzhou (Zhejiang)	38	24	Huainan (Anhui)	32
10	Wuhu (Anhui)	38	25	Guiyang (Guizhou)	32
11	Yunfo (Guangdong)	37	26	Lu'an (Anhui)	31
12	Heyuan (Guangdong)	37	27	Qingyuan (Guangdong)	31
13	Huaibei (Anhui)	37	28	Tianjin	31
14	Qingdao (Shandong)	36	29	Shantou (Guangdong)	30
15	Zunyi (Guizhou)	35	30	Zhongshan (Guangdong)	30

solicit public opinions, enhance public understanding of the operation of government agencies, and communicate effectively the relations between the public and government agencies.

With the arrival of the information age and the popularity of mobile Internet technology, more and more city management departments publish related information through the Internet public account; more and more people express their opinions and participate in the decision-making through the Internet platform. Online public participating in the questionnaire survey, program design, etc., is becoming an indispensable part of the intelligent city construction.

This book selects Sina micro-blog and Twitter site as two online platforms for domestic and foreign intelligent city evaluation. In 2006, Twitter website launched the world's first micro blog service, which quickly became popular after the U.S. presidential campaign. In 2007, the domestic emergence of a large number of followers of Twitter, but these early local service providers are lack of experience, simply imitating foreign products, but with less user usage and less attention. In August 2009, Sina launched Sina MicroBlog, and invited many celebrities to enter. Sina developed the nickname of "Weibo" into the well-known Internet buzzword in just half years by right of celebrity effect, successfully obtained the leading position in the domestic MicroBlog market. In 2010, MicroBlog appeared blowout growth, the major portals, government websites, media units and life service websites have been launched in succession, and the first year of China MicroBlog opens. In the next two years in 2011 and in 2012, the domestic MicroBlog user group continued to grow. Government, schools, merchants and other institutions have launched the official MicroBlog, and people have become accustomed to concern, inquiry, register, participate and feedback all kind of activities through MicroBlog.

This book selected the proportion of the number of followers of intelligent city official accountant (for domestic cities the data of Sina MicroBlog were adopted, while for foreign cities the data of Twitter were adopted) to the total city population as the evaluation of Online public participation ratio.

- Indicator F23: Residents' health electronic archives usage level

The indicator refers to the proportion of the number of residents having personal health electronic records in city's total number of residents, reflecting the digitization degree of citizen information.

Intelligent health care is an important part of intelligent city. Information technology such as Internet of Things shall be utilized to achieve interaction between patients and medical staffs, medical institutions, medical equipment, and gradually achieve digitization, informatization through the construction of health records medical information platform. The applications of intelligent health care in the intelligent city are not only for meeting the needs of seeing doctor inside the hospital, but should also include the application and promotion of public health system in the city area, which is the future development direction of intelligent health care.

In the intelligent public health system, Electronic Health Records (EHRs) are electronic records with preservation and reference value which are accumulated in people's related health activities, stored in the computer systems, providing services and a life-long personal health records with security. The residents' personal health is the core of electronic health records, through the whole life cycle, covering all relevant factors of health, realizing multi-channel dynamic information collection, and meeting the needs of information resources for residents' self-health care, health management and health decision-making (Dong 2010).

This book uses a gradient evaluation method in the calculation, with 100 (excellent level) indicating that there's a detailed electronic health records application (such as related applications of public health guidance, self-searching medical treatment guidance, medical insurance, citizen case networking, telemedicine system on city website), with 50 (average), indicating that that the application is without specific content or only used in part of the region, and with 0 indicating totally no application.

- Indicator F24: Level of intelligent coping with emergency

The indicator refers to the level of intelligent emergency system in the face of major urban emergencies (such as disasters, accidents, etc.).

City safety is an typical public safety, the purpose is to protect public health, life and property from damage, to control all kinds of threats at a minimum degree in the control way of socialization and legalization, and to maintain the normal order of public life as stable as possible (Cai 2012). In the intelligent city construction, the way to use information technology to develop a set of intelligent emergency system and working mechanism integrating prevention and emergency preparedness, monitoring and early warning, emergency response and rescue, can reflect the real-time response and learning ability of intelligent city.

This book uses a gradient evaluation method in the calculation, with 100 (excellent level) indicating that there's a detailed electronic health records application (such as related applications of public health guidance, self-searching medical treatment guidance, medical insurance, citizen case networking, telemedicine system on city website), with 50 (average), indicating that that the application is without specific content or only used in part of the region, and with 0 indicating totally no application.

A3.3 Intelligent Smart Economy and Industry

- Indicator F31: The proportion of R&D expenditure in GDP

This indicator refers to the proportion of city R&D expenditures accounted for GDP, reflecting the city's scientific and technological strength, innovation and core competitiveness.

R&D expenditures refer to expenditures on research and development in the whole society, including expenditures for basic research, applied research and experimental development. There is a significant correlation between the input of city R&D and the number of scientific research personnel, and the number of enterprises engaged in scientific research in the city, reflecting the city's scientific and technological strength, innovation and core competitiveness. In the construction of intelligent city, R&D expenditures have become an important indicator to judge the ability of city's independent innovation ability.

This book adopts the data of science and technology expenditures in item 2-24 in "China Urban Statistical Yearbook 2011" in the evaluation of domestic intelligent city, and adopts the data of R&D expenditures of the World Bank accounted for GDP in the evaluation of international intelligent city.[8]

---

[8]The weight of the Word Bank's R&D expenditures accounted for GDP is the national data. The city's data can only be replaced by the country's data in the evaluation of intelligent city's construction.

- Indicator F32: City labor productivity

This indicator refers to the per capita gross domestic product (GDP), reflecting the city's development level of intellectual economic.

- Indicator F33: City product value density

The indicator refers to the average value of GNP created by the cities per square kilometer of land, which fully reflects the intelligence level efficiency of land use.

The city labor productivity and the density of the city output value are the effective reflection of the city's economic development. The urban labor productivity is more objective to measure the living standard of the people of all countries, and the density of city output value more objectively reflects land efficiency.

In the evaluation of the two indicators above, this book selected the data of GNP/population and GNP/city, respectively.

- Indicator F34: The proportion of city intelligent industry

This indicator refers to the proportion of knowledge and technology intensive industries in urban industries.

Intelligent industry is an advanced stage of industrial development, an important direction of transformation and upgrading of traditional industries. The construction of Intelligent City takes the Internet of Things, cloud computing, mobile Internet, big data and other smart industries as the technological basis, promoting industrial development in the fields of information management services, information technology related manufacturing, maintenance and design of information system, information analysis and consulting firstly, and on this basis, impacting on the wider areas of city public management and innovation service. The construction of intelligent city will introduce the information digital technology into modern manufacturing industry and service industry, generating the "Intelligent Industry" different from the traditional.

This book will take the proportion of e-commerce transactions and the city GDP as the judge of the proportion of intelligent industry, and for the one without city data will alternatively adopt the data of the country or province.

A3.4 Smart Hardware Facility

- Indicator F41: Public space free network coverage density

This indicator refers to proportion of the city space offering free wireless network in the total area of the city, reflecting the city's information access level from the hardware's.

The coverage level of city network is one of the important indicators to measure the city's intelligence construction. In 2004, the Philadelphia put forward the concept of "wireless city", and built a wireless broadband metropolitan area network based on WLAN standards. And then a number of cities around the world have begun to invest in the construction of wireless city, based on high-speed

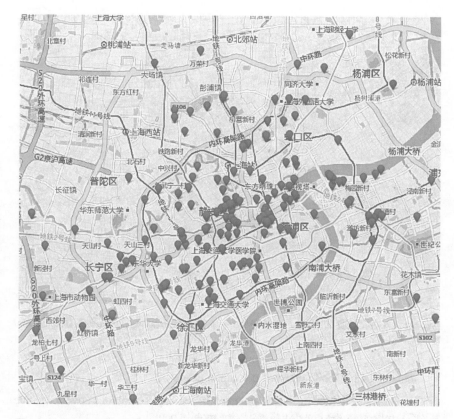

Fig. A.1 Shanghai local Wi-Fi free coverage area. *Picture source* http://wlan.vnet.cn

broadband wireless network, to achieve the goal of accessing the wireless network and information services at anytime and anywhere. In 2013, China issued the "Broadband China" strategy and Implementation Plan. By the end of 2013, the WLAN to achieve the goal of accessing the city's major public hotspots.[9]

This book selected China Telecom Wi-Fi Hot Query Platform[10] to evaluate the providing level of domestic intelligent city's public space free network, and selected Free Wi-Fi Hot Query Platform[11] to evaluate the providing level of abroad intelligent city's public space free network. These two platforms provide Wi-Fi hotspots, and the quotient of the number and the area of the city is used as an indicator to evaluate the coverage density of the free network. Wi-Fi free coverage areas in Shanghai and London are shown in Figs. A.1 and A.2, respectively.

---

[9]For details see http://www.gov.cn/zwgk/2013-08/17/content_2468348.htm.
[10]For details see http://wlan.vnet.cn.
[11]For details see http://www.wificafespots.com/wifi.

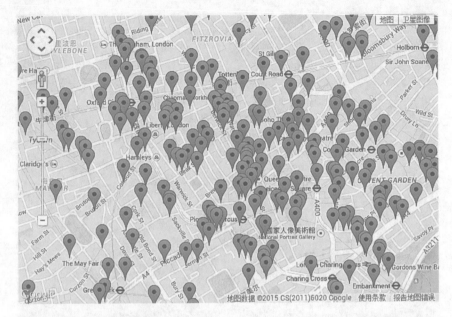

Fig. A.2 London local Wi-Fi free coverage area. *Picture source* http://www.wificafespots.com/
wifi/city/GB–City_of_London

- Indicator F42: Mobile network per capita usage

This indicator refers to the usage rate of per capita mobile network (phone
3G/4G), reflecting the construction of city mobile network.

Similar to the free network coverage density of public space, the per capita
utilization rate of mobile network is also the evaluation of the construction level of
city Internet. The former is more from the perspective of the government to provide
infrastructure, while the latter evaluates the city's mobile Internet's usage level form
the perspective of popularity rate of mobile network facilities, such as mobile
phone.

This book evaluates the per capita utilization of city mobile networks through
the ratio of the number of people using the city's mobile network.

- Indicator F43: City broadband speed

This indicator is one of the basic links in the construction of intelligent city. If a city
proposed the construction of intelligent city, while its network speed is very slow, so
naturally, there will be doubts about the construction of intelligent city. NetIndex
provides the real measurement information about the world city broadband speed.[12]

---

[12]Data source: http://www.netindex.com.

- Indicator F44: intelligent grid level of coverage

The indicator refers to the intelligent power grid coverage in the city, reflecting the city's intelligence level of energy.

Intelligent power grid is the intelligence of power grid. Compared with the traditional power grid, the advanced nature of the intelligent power grid is mainly reflected in the link of matching point. City intelligent power combines an advanced sensor technology, information communication technology, analysis and decision technology, automatic control technology and energy power technology, and highly integrates with city power grid infrastructure, to form of new modern city power grid.

The book adopts gradient evaluation method in the calculation, with 100 (excellent level) indicating smart power grid application with details, with 50 (average), indicating that that the application is without specific content or only used in part of the region, and with 0 indicating totally no application.

A3.5 Residents' Intelligent Potential

- Indicator F51: Proportion of city netizens

The indicator refers to the proportion of netizens in city population, reflecting the level of accessing information and learning.

With the development and popularization of computer and network, the number of people in the world is increasing, and the network has become an important platform to reflect public opinion and convey the voice of the people. The Internet is gradually shifting and replacing traditional media, playing a more and more important role in people's production and living.

Intelligent city should have the qualification of accessing to information, sending information, and immediate feeding back information through the network platform. The proportion of city netizens can reflect the degree of intelligence of the city from the perspective of the citizen.

This book adopts the data of number of Internet users in item 2-36 in "China Urban Statistical Yearbook 2011" in the evaluation of domestic intelligent city, and adopts the data of the World Bank Internet usage ratio in the evaluation of intelligent city abroad.[13]

- Indicator F52: Proportion of information practitioners

The indicator refers to the proportion of information practitioners in all the city employees.

The proportion of information practitioners can reflect the proportion of the city's information services and software industry, the citizen demands for information technology services, as well as the city's innovation capacity. This is an important aspect in evaluating intelligent city.

---

[13]The number of Internet users in the World Bank is the national data. The city's data can only be replaced by the country's data in the evaluation of intelligent city's construction.

This book adopts the data of information transmission, computer services and software industry practitioners in item 2-7 in "China Urban Statistical Yearbook 2011" in the evaluation of domestic intelligent city, the data reported in "The European CTC lusters" in the evaluation of European intelligent city,[14] and the statistics of the number of employees of US Department of Labor in the evaluation of American intelligent city.[15]

- Indicator F53: Proportion of college degree or above

The indicator refers to the proportion of the population with college degree or above accounted for the city's total population, reflecting city's intelligence level through educational level.

The education level of the population is an important indicator to measure the cultural quality of the population, and the proportion of city residents with higher education can reflect the degree of city's development.

This book adopts the data of number of students in higher education in item 2-29 in "China Urban Statistical Yearbook 2011" in the evaluation of domestic intelligent city, and adopts education rankings published by Organization for Economic Co-operation and Development (OECD) in the evaluation of intelligent city abroad.[16]

- Indicator F54: number of citizens online spending per capita

The indicator refers to the proportion of the amount of per capita net consumption in the total consumption amount, indirectly reflecting the popularity of the Internet and the development level of the Internet of Things.

Research from Zhao (2009) shows that turnover generated by European e-commerce has accounted for 1/4 of total business, while in the United States it has been up to 1/3 above. Well-known e-commerce companies, such as American Online, YAHOO, E-bay, began to rise around 1995, and IBM, Amazon, Wal-Mart supermarket and other e-commerce companies have made a huge profit in their respective areas. In China, the prosperity and development of the online shopping market has greatly stimulated the city's economic growth, and online shopping grows fastest in the retail sales of social consumer, which has become the new driving force of economic growth.

This book adopts city online data in "China's City Online Shopping Development Environmental Statement" published by Taobao in 2012 (the book will use 0, 25, 50, 75 and 100 to evaluate the city, respectively) in the evaluation of

---

[14]For details see: http://rucforsk.ruc.dk/site/files/32956338/the_european_ict_clusters_web_0.pdf.
[15]For details see: http://www.bls.gov/oes.
[16]Education rankings published by Organization for Economic Co-operation and Development (OECD) are the national data. The city's data can only be replaced by the country's data in the evaluation of intelligent city's construction.

domestic intelligent city, and adopts the data in "Global Perspective on Retail: Online Retailing" published by Cushman & Wakefield in 2003 in the evaluation of international intelligent city.[17]

A4 Data processing of Intelligent City Evaluation Indicator

The dimensions of the original data are different, not only including physical quantity, value, but also including the per capita value, percentage, so they are not able to be directly incorporated into the evaluation indicator system for comparison. To cope with the disparity of dimensions of each indicator to carry out comprehensive summary, after the data collection work, it's also needed to do the standardization processing of the original data, making it into a dimensionless numerical value, to eliminate the influence of different computing units, to stabilize the data.

The influence of dimension can be eliminated by selecting simple and practical method. The main principle is to determine a comparison standard for the indicator to be evaluated firstly, as the compared standard value, and compare the actual value and relative value of each indicator, and then it can convert a variety of indicators in different natures and measurements to the same measurement indicators.

There are many standardization methods for indicators, such as linear, fold line and curve.

The linear method assumes a linear relationship between the actual value and the normalized value of the indicators. There are two common ways to deal with:

(1) Centralized method:

$$A'_i = A_i - \bar{A}$$

(2) Standardization method:

$$A'_i = (A_i - \bar{A})/\sigma_i, \sigma_i^2 = \sum(A_i - \bar{A})/n^2$$

or

$$A'_i = (A_i - A_{\min})/A_{\max} - A_{\min}$$

---

[17]Data released by the Cushman & Wakefield is the national data. The city's data can only be replaced by the country's data in the evaluation of intelligent city's construction.

or

$$A'_i = A_i/A_{max}$$

The fold line method is mainly used in the evaluation of the overall level of things affected by different interval changes, to use the extreme method to standardize the treatment in segment. However, if the influence of the actual value to the value of the evaluation is not equal, then curve type's standardization method shall be used. Here, we adopt $A'_i = A_i/A_{max}$, so the value of each indicator is in the range of 0–100 after conversion, which is in accordance with the centennial grading system.

A5 An Overview on the Basic Data of Construction Level Ranking of Intelligent Cities in China and in the World

See Tables A.16, A.17, A.18, A.19, A.20, A.21, A.22, A.23, A.24, A.25, A.26 and A.27.

Table A.16 Intelligent construction comprehensive evaluation and divided evaluation of 33 cities in China

City	Overall		Intelligent environment and construction		Intelligent management and services		Intelligent economy and industry		Intelligent hardware facilities		Residents' intelligent potential	
	Ranking	Score	Ranking	Score	Ranking	Score	Ranking	Score	Ranking	Score	Ranking	Score
Jinhua	1	62.92	3	86.08	28	36.50	3	56.17	1	77.91	1	57.93
Ningbo	2	57.09	6	82.65	3	70.50	23	32.29	6	58.79	8	41.20
Zhuhai	3	56.13	14	71.31	7	61.47	27	27.70	2	69.84	4	50.33
Wenzhou	4	55.80	8	81.14	26	39.98	20	33.60	3	68.58	2	55.67
Wuhan	5	55.44	9	79.84	4	69.29	24	31.18	9	54.55	7	42.33
Nanjing	6	54.78	18	66.35	2	74.94	16	35.16	5	59.02	9	38.45
Wuxi	7	54.67	7	81.60	11	60.22	2	57.65	14	46.50	22	27.41
Pudong, Shanghai	8	54.48	11	75.45	1	75.65	11	40.45	13	50.61	18	30.24
Taizhou	9	53.98	19	66.19	13	58.66	1	66.51	18	42.83	11	35.72
Changzhou	10	53.25	16	70.37	21	46.38	4	52.27	4	64.39	15	32.84
Weihai	11	53.22	2	87.11	14	55.57	9	43.67	16	44.75	13	34.99
Zhenjiang	12	53.09	5	84.55	22	46.16	5	51.68	12	51.60	16	31.46
Dongying	13	51.97	1	87.50	10	60.28	7	48.44	27	28.44	12	35.19
Langfang	14	51.85	4	84.98	32	23.16	13	40.11	8	56.16	3	54.81
Dezhou	15	48.41	13	73.02	9	60.55	8	44.31	19	41.25	24	22.90
Xianyang	16	47.65	10	78.78	16	50.41	18	34.43	23	36.95	10	37.66
Ya'an	17	45.77	23	50.00	24	42.49	15	35.16	10	53.11	6	48.09
Nanping	18	45.47	33	25.00	18	48.96	6	51.64	11	52.45	5	49.29
Zhuzhou	19	43.61	21	63.57	12	59.59	25	29.91	22	37.01	21	27.95
Tongling	20	42.75	24	50.00	25	40.34	14	37.28	7	56.22	19	29.91
Wuhu	21	42.69	22	57.49	19	46.55	12	40.26	20	40.73	20	28.42
Lasa	22	40.82	17	67.33	17	49.49	33	7.90	17	44.58	14	34.78

(continued)

Table A.16 (continued)

City	Overall		Intelligent environment and construction		Intelligent management and services		Intelligent economy and industry		Intelligent hardware facilities		Residents' intelligent potential	
	Ranking	Score	Ranking	Score	Ranking	Score	Ranking	Score	Ranking	Score	Ranking	Score
Changzhi	23	40.75	20	63.78	6	63.67	28	25.13	26	30.92	26	20.25
Bengbu	24	40.43	25	50.00	8	61.23	17	34.72	21	38.58	27	17.63
Huainan	25	40.36	12	75.00	23	44.05	22	32.41	24	35.65	29	14.67
Pingxiang	26	38.67	26	50.00	29	34.62	10	43.18	15	44.78	25	20.76
Hebi	27	37.57	27	50.00	5	67.73	26	28.72	28	27.88	30	13.55
Qinhuangdao	28	36.60	30	49.33	15	55.02	30	20.53	29	27.47	17	30.67
Handan	29	35.19	15	70.97	33	20.79	19	34.25	30	23.88	23	26.06
Liupanshui	30	33.03	28	50.00	20	46.48	29	22.92	25	34.81	33	10.92
Luohe	31	30.22	29	50.00	31	31.14	21	33.35	31	19.83	28	16.77
Wuhai	32	22.48	31	37.50	27	38.93	32	16.32	33	6.91	31	12.72
Liaoyuan	33	21.55	32	37.50	30	32.50	31	17.56	32	9.23	32	10.95

Table A.17 Intelligent environment and construction original data and score of 33 cities in China

Ranking	City	City PM2.5/PM10 monitoring stations density		City grid management level of coverage		Residents' intelligent transportation tools usage level		Online publishing level of city future construction plan		Synthetical value
		Original data a	Score	Original data b	Score	Original data b	Score	Original data b	Score	
1	Dongying	0.0918	100.00	100.00	100.00	100.00	100.00	50.00	50.00	87.50
2	Weihai	0.0444	48.44	100.00	100.00	100.00	100.00	100.00	100.00	87.11
3	Jinhua	0.0407	44.34	100.00	100.00	100.00	100.00	100.00	100.00	86.08
4	Langfang	0.0825	89.92	100.00	100.00	100.00	100.00	50.00	50.00	84.98
5	Zhenjiang	0.0351	38.21	100.00	100.00	100.00	100.00	100.00	100.00	84.55
6	Ningbo	0.0281	30.60	100.00	100.00	100.00	100.00	100.00	100.00	82.65
7	Wuxi	0.0242	26.39	100.00	100.00	100.00	100.00	100.00	100.00	81.60
8	Wenzhou	0.0684	74.57	50.00	50.00	100.00	100.00	100.00	100.00	81.14
9	Wuhan	0.0178	19.37	100.00	100.00	100.00	100.00	100.00	100.00	79.84
10	Xianyang	0.0597	65.11	50.00	50.00	100.00	100.00	100.00	100.00	78.78
11	Pudong, Shanghai	0.0017	1.80	100.00	100.00	100.00	100.00	100.00	100.00	75.45
12	Huainan	0.0000	0.00	100.00	100.00	100.00	100.00	100.00	100.00	75.00
13	Dezhou	0.0845	92.10	50.00	50.00	100.00	100.00	50.00	50.00	73.02
14	Zhuhai	0.0324	35.26	50.00	50.00	100.00	100.00	100.00	100.00	71.31
15	Handan	0.0770	83.88	50.00	50.00	100.00	100.00	50.00	50.00	70.97
16	Changzhou	0.0289	31.47	50.00	50.00	100.00	100.00	100.00	100.00	70.37
17	Lasa	0.0636	69.33	100.00	100.00	50.00	50.00	50.00	50.00	67.33
18	Nanjing	0.0141	15.38	50.00	50.00	100.00	100.00	100.00	100.00	66.35
19	Taizhou	0.0594	64.77	50.00	50.00	50.00	50.00	100.00	100.00	66.19
20	Changzhi	0.0506	55.13	100.00	100.00	50.00	50.00	50.00	50.00	63.78

(continued)

Table A.17 (continued)

Ranking	City	City PM2.5/PM10 monitoring stations density		City grid management level of coverage		Residents' intelligent transportation tools usage level		Online publishing level of city future construction plan		Synthetical value
		Original data a	Score	Original data b	Score	Original data b	Score	Original data b	Score	
21	Zhuzhou	0.0498	54.30	50.00	50.00	100.00	100.00	50.00	50.00	63.57
22	Wuhu	0.0275	29.96	100.00	100.00	50.00	50.00	50.00	50.00	57.49
23	Luohe	0.0000	0.00	50.00	50.00	100.00	100.00	50.00	50.00	50.00
24	Bengbu	0.0000	0.00	50.00	50.00	100.00	100.00	50.00	50.00	50.00
25	Tongling	0.0000	0.00	50.00	50.00	50.00	50.00	100.00	100.00	50.00
26	Ya'an	0.0000	0.00	50.00	50.00	50.00	50.00	100.00	100.00	50.00
27	Pingxiang	0.0000	0.00	100.00	100.00	0.00	0.00	100.00	100.00	50.00
28	Hebi	0.0000	0.00	50.00	50.00	50.00	50.00	100.00	100.00	50.00
29	Liupanshui	0.0000	0.00	50.00	50.00	100.00	100.00	50.00	50.00	50.00
30	Qinhuangdao	0.0434	47.32	50.00	50.00	50.00	50.00	50.00	50.00	49.33
31	Wuhai	0.0000	0.00	50.00	50.00	50.00	50.00	50.00	50.00	37.50
32	Liaoyuan	0.0000	0.00	50.00	50.00	0.00	0.00	100.00	100.00	37.50
33	Nanping	0.0000	0.00	50.00	50.00	0.00	0.00	50.00	50.00	25.00

Unit of a: pcs/km^2
Unit of b: none

Table A.18 Intelligent management and service original data and score of 33 cities in China

Ranking	City	Online publicity level of the government's non-classified documents		Online public participation ratio		Residents' health electronic archives usage level		Emergency intelligent level of emergency		Synthetical value
		Original data a	Score	Original data b	Score	Original data a	Score	Original data a	Score	
1	Pudong, Shanghai	45	100.00	2.84	2.59	100.00	100.00	100.00	100.00	75.65
2	Nanjing	21	46.99	0.5779	52.78	100.00	100.00	100.00	100.00	74.94
3	Ningbo	23	50.35	0.3465	31.64	100.00	100.00	100.00	100.00	70.50
4	Wuhan	23	51.39	0.2822	25.77	100.00	100.00	100.00	100.00	69.29
5	Hebi	18	40.96	0.3278	29.94	100.00	100.00	100.00	100.00	67.73
6	Changzhi	43	95.45	0.1012	9.24	50.00	50.00	100.00	100.00	63.67
7	Zhuhai	41	91.07	0.0528	4.83	50.00	50.00	100.00	100.00	61.47
8	Bengbu	19	42.23	0.0296	2.70	100.00	100.00	100.00	100.00	61.23
9	Dezhou	19	42.20	1.0949	100.00	50.00	50.00	50.00	50.00	60.55
10	Dongying	14	30.76	0.1133	10.35	100.00	100.00	100.00	100.00	60.28
11	Wuxi	22	48.42	0.4648	42.45	50.00	50.00	100.00	100.00	60.22
12	Zhuzhou	13	29.34	0.0988	9.02	100.00	100.00	100.00	100.00	59.59
13	Taizhou	18	39.80	1.0383	94.83	50.00	50.00	50.00	50.00	58.66
14	Weihai	16	35.22	0.4060	37.08	50.00	50.00	100.00	100.00	55.57
15	Qinhuangdao	18	40.07	0.3285	30.00	100.00	100.00	50.00	50.00	55.02
16	Xianyang	22	50.26	0.0149	1.36	100.00	100.00	50.00	50.00	50.41
17	Lasa	4	8.39	0.4335	39.59	100.00	100.00	50.00	50.00	49.49
18	Nanping	19	41.67	0.0458	4.18	100.00	100.00	50.00	50.00	48.96
19	Wuhu	38	84.62	0.0172	1.57	100.00	100.00	0.00	0.00	46.55

(continued)

Appendix

Table A.18 (continued)

Ranking	City	Online publicity level of the government's non-classified documents		Online public participation ratio		Residents' health electronic archives usage level		Emergency intelligent level of emergency		Synthetical value
		Original data a	Score	Original data b	Score	Original data a	Score	Original data a	Score	
20	Liupanshui	16	35.94	0.0000	0.00	100.00	100.00	50.00	50.00	46.48
21	Changzhou	16	35.40	0.0014	0.13	100.00	100.00	50.00	50.00	46.38
22	Zhenjiang	21	47.54	0.4060	37.08	100.00	100.00	0.00	0.00	46.16
23	Huainan	32	70.80	0.0592	5.41	100.00	100.00	0.00	0.00	44.05
24	Ya'an	4	8.39	0.1267	11.57	100.00	100.00	50.00	50.00	42.49
25	Tongling	22	48.46	0.1410	12.88	50.00	50.00	50.00	50.00	40.34
26	Wenzhou	20	44.52	0.1688	15.41	50.00	50.00	50.00	50.00	39.98
27	Wuhai	3	5.59	0.0013	0.12	100.00	100.00	50.00	50.00	38.93
28	Jinhua	18	39.29	0.0732	6.69	100.00	100.00	0.00	0.00	36.50
29	Pingxiang	23	52.12	0.3980	36.35	0.00	0.00	50.00	50.00	34.62
30	Liaoyuan	10	22.37	0.0835	7.63	50.00	50.00	50.00	50.00	32.50
31	Luohe	10	22.37	0.0240	2.19	100.00	100.00	0.00	0.00	31.14
32	Langfang	14	31.32	0.1241	11.33	50.00	50.00	0.00	0.00	23.16
33	Handan	13	27.97	0.0567	5.18	50.00	50.00	0.00	0.00	20.79

Unit of a: none
Unit of b: %

Table A.19 Intelligent economy and industry original data and score of 33 cities in China

Ranking	City	The proportion of R&D expenditure in GDP		City labor productivity		City product value density		The proportion of city intelligent industry		Synthetical value
		Original data a	Score	Original data b	Score	Original data c	Score	Original data a	Score	
1	Taizhou	0.10	3.35	323948	71.79	297474	100.00	3.75	90.89	66.51
2	Wuxi	0.20	6.86	297341	65.89	199170	66.95	3.75	90.89	57.65
3	Jinhua	0.07	2.42	451249	100.00	284066	95.49	1.11	26.77	56.17
4	Changzhou	0.16	5.63	247176	54.78	171874	57.78	3.75	90.89	52.27
5	Zhenjiang	0.10	3.29	247877	54.93	171429	57.63	3.75	90.89	51.68
6	Nanping	0.05	1.58	375517	83.22	257984	86.73	1.45	35.02	51.64
7	Dongying	0.05	1.61	383232	84.93	216548	72.80	1.42	34.41	48.44
8	Dezhou	0.01	0.20	280297	62.12	239437	80.49	1.42	34.41	44.31
9	Weihai	0.10	3.54	398504	88.31	144052	48.43	1.42	34.41	43.67
10	Pingxiang	0.05	1.70	138261	30.64	120092	40.37	4.13	100.00	43.18
11	Pudong, Shanghai	2.90	100.00	105971	23.48	45307	15.23	0.95	23.10	40.45
12	Wuhu	0.64	22.20	99484	22.05	74227	24.95	3.79	91.84	40.26
13	Langfang	0.05	1.60	288889	64.02	214521	72.11	0.94	22.72	40.11
14	Tongling	0.16	5.35	119867	26.56	75444	25.36	3.79	91.84	37.28
15	Ya'an	0.02	0.52	151731	33.62	152381	51.22	2.28	55.28	35.16
16	Nanjing	0.30	10.46	76693	17.00	66331	22.30	3.75	90.89	35.16
17	Bengbu	0.33	11.42	76913	17.04	55300	18.59	3.79	91.84	34.72
18	Xianyang	0.03	0.91	133093	29.49	164105	55.17	2.16	52.16	34.43
19	Handan	0.03	1.06	154674	34.28	196699	66.12	1.47	35.53	34.25

(continued)

Table A.19 (continued)

Ranking	City	The proportion of R&D expenditure in GDP		City labor productivity		City product value density		The proportion of city intelligent industry		Synthetical value
		Original data a	Score	Original data b	Score	Original data c	Score	Original data a	Score	
20	Wenzhou	0.11	3.74	235318	52.15	153947	51.75	1.11	26.77	33.60
21	Luohe	0.04	1.25	123394	27.35	113667	38.21	2.75	66.61	33.35
22	Huainan	0.12	4.21	57259	12.69	62161	20.90	3.79	91.84	32.41
23	Ningbo	0.25	8.64	192631	42.69	151943	51.08	0.0111	26.77	32.29
24	Wuhan	0.17	5.95	99418	22.03	108902	36.61	0.0248	60.14	31.18
25	Zhuzhou	0.15	5.28	136739	30.30	105796	35.56	0.0200	48.48	29.91
26	Hebi	0.05	1.68	101205	22.43	71856	24.16	0.0275	66.61	28.72
27	Zhuhai	0.33	11.42	160717	35.62	97218	32.68	0.0128	31.09	27.70
28	Changzhi	0.06	1.97	137594	30.49	155182	52.17	0.0066	15.88	25.13
29	Liupanshui	0.04	1.52	173472	38.44	130036	43.71	0.0033	8.01	22.92
30	Qinhuangdao	0.05	1.82	106651	23.63	100955	33.94	0.0094	22.72	20.53
31	Liaoyuan	0.05	1.73	82661	18.32	88476	29.74	0.0085	20.45	17.56
32	Wuhai	0.15	5.26	88409	19.59	61825	20.78	0.0081	19.66	16.32
33	Lasa	0.00	0.00	83214	18.44	28453	9.56	0.0015	3.60	7.90

Unit of a: %
Unit of b: Yuan/person
Unit of c: ten thousand Yuan/km^2

Table A.20 Intelligent hardware facilities original data and score of 33 cities in China

Ranking	City	Public space free network coverage density		Mobile network per capita usage		City broadband speed		Intelligent grid level of coverage		Synthetical value
		Original data a	Score	Original data b	Score	Original data c	Score	Original data d	Score	
1	Jinhua	16.07	100.00	103.40	100.00	29.67	61.66	50.00	50.00	77.91
2	Zhuhai	3.81	23.71	96.27	93.10	30.10	62.55	100.00	100.00	69.84
3	Wenzhou	4.38	27.28	77.23	74.69	34.81	72.34	100.00	100.00	68.58
4	Changzhou	8.42	52.40	39.87	38.55	32.05	66.60	100.00	100.00	64.39
5	Nanjing	3.51	21.84	32.64	31.56	39.78	82.67	100.00	100.00	59.02
6	Ningbo	4.32	26.91	37.18	35.96	35.19	73.13	100.00	100.00	58.79
7	Tongling	6.62	41.19	37.38	36.15	27.83	57.83	100.00	100.00	56.22
8	Langfang	3.13	19.50	30.80	29.79	36.38	75.60	100.00	100.00	56.16
9	Wuhan	0.76	4.72	53.33	51.58	32.89	68.35	100.00	100.00	54.55
10	Ya'an	3.23	20.09	45.43	43.93	26.07	54.18	50.00	50.00	53.11
11	Nanping	1.19	7.41	56.90	55.03	48.12	100.00	50.00	50.00	52.45
12	Zhenjiang	5.04	31.36	77.48	74.93	25.76	53.53	100.00	100.00	51.60
13	Pudong, Shanghai	5.00	31.14	30.41	29.41	22.07	45.86	100.00	100.00	50.61
14	Wuxi	0.83	5.14	20.87	20.18	37.10	77.10	50.00	50.00	46.50
15	Pingxiang	4.36	27.16	31.91	30.86	34.22	71.11	50.00	50.00	44.78
16	Weihai	0.56	3.50	49.18	47.56	37.50	77.93	50.00	50.00	44.75
17	Lasa	4.76	29.59	27.91	26.99	10.47	21.76	100.00	100.00	44.58
18	Taizhou	6.40	39.85	19.42	18.78	30.16	62.68	50.00	50.00	42.83
19	Dezhou	0.24	1.49	19.79	19.13	21.36	44.39	100.00	100.00	41.25

(continued)

Table A.20 (continued)

Ranking	City	Public space free network coverage density		Mobile network per capita usage		City broadband speed		Intelligent grid level of coverage		Synthetical value
		Original data a	Score	Original data b	Score	Original data c	Score	Original data d	Score	
20	Wuhu	4.32	26.90	33.16	32.07	25.96	53.95	50.00	50.00	40.73
21	Bengbu	2.06	12.79	15.38	14.88	36.89	76.66	50.00	50.00	38.58
22	Zhuzhou	1.79	11.16	25.76	24.91	29.82	61.97	50.00	50.00	37.01
23	Xianyang	2.54	15.80	29.07	28.12	25.92	53.87	50.00	50.00	36.95
24	Huainan	7.74	48.18	11.32	10.95	40.16	83.46	0.00	0.00	35.65
25	Liupanshui	0.94	5.82	41.58	40.21	20.80	43.23	50.00	50.00	34.81
26	Changzhi	0.42	2.62	35.89	34.70	17.49	36.35	50.00	50.00	30.92
27	Dongying	0.08	0.51	19.49	18.85	21.36	44.39	50.00	50.00	28.44
28	Hebi	0.26	1.60	2.41	2.33	27.72	57.61	50.00	50.00	27.88
29	Qinhuangdao	0.60	3.72	13.76	13.31	20.63	42.87	50.00	50.00	27.47
30	Handan	0.05	0.32	16.14	15.61	14.24	29.59	50.00	50.00	23.88
31	Luohe	0.12	0.73	21.71	21.00	27.72	57.61	0.00	0.00	19.83
32	Liaoyuan	0.22	1.34	2.02	1.95	16.19	33.65	0.00	0.00	9.23
33	Wuhai	0.24	1.48	4.55	4.40	10.47	21.76	0.00	0.00	6.91

Unit of a: pcs/km^2

Unit of b: %

Unit of c: Mbps

Unit of d: none

Table A.21 Residents' intelligent potential original data and score of 33 cities in China

Ranking	City	The proportion of Internet users in city		The proportion of information professionals		The proportion of junior college or above educational level population		Residents' per capita online shopping expenditure amount		Synthetical value
		Original data a	Score	Original data a	Score	Original data a	Score	Original data b	Score	
1	Jinhua	163.33	46.18	1.27	66.50	16.25	69.04	50.00	50.00	57.93
2	Wenzhou	353.66	100.00	0.43	22.31	5.97	25.38	75.00	75.00	55.67
3	Langfang	89.09	25.19	0.84	44.18	23.50	99.86	50.00	50.00	54.81
4	Zhuhai	39.84	11.27	1.10	57.37	13.58	57.70	75.00	75.00	50.33
5	Nanping	96.52	27.29	1.91	100.00	10.56	44.87	25.00	25.00	49.29
6	Ya'an	36.44	10.30	1.09	57.06	23.53	100.00	25.00	25.00	48.09
7	Wuhan	29.92	8.46	0.40	21.03	15.26	64.82	75.00	75.00	42.33
8	Ningbo	65.55	18.53	0.40	20.72	6.01	25.54	100.00	100.00	41.20
9	Nanjing	25.76	7.28	0.23	11.95	14.02	59.58	75.00	75.00	38.45
10	Xianyang	26.20	7.41	0.41	21.55	10.99	46.70	75.00	75.00	37.66
11	Taizhou	63.30	17.90	0.86	44.87	7.09	30.13	50.00	50.00	35.72
12	Dongying	44.92	12.70	0.81	42.48	8.37	35.58	50.00	50.00	35.19
13	Weihai	65.76	18.59	0.35	18.23	12.51	53.14	50.00	50.00	34.99
14	Lasa	0.000	0.00	1.72	90.05	5.66	24.07	25.00	25.00	34.78
15	Changzhou	68.13	19.27	0.34	17.82	10.42	44.28	50.00	50.00	32.84
16	Zhenjiang	46.99	13.29	0.30	15.91	10.98	46.65	50.00	50.00	31.46
17	Qinhuangdao	45.51	12.87	0.32	16.80	10.12	42.99	50.00	50.00	30.67
18	Pudong, Shanghai	78.25	22.13	0.28	14.53	2.18	9.28	75.00	75.00	30.24
19	Tongling	24.65	6.97	0.21	10.74	6.33	26.92	75.00	75.00	29.91

(continued)

Table A.21 (continued)

Ranking	City	The proportion of Internet users in city		The proportion of information professionals		The proportion of junior college or above educational level population		Residents' per capita online shopping expenditure amount		Synthetical value
		Original data a	Score	Original data a	Score	Original data a	Score	Original data b	Score	
20	Wuhu	22.64	6.40	0.17	8.68	11.44	48.61	50.00	50.00	28.42
21	Zhuzhou	19.88	5.62	0.43	22.46	7.93	33.71	50.00	50.00	27.95
22	Wuxi	60.43	17.09	0.34	17.83	5.82	24.73	50.00	50.00	27.41
23	Handan	67.19	19.00	0.35	18.30	3.98	16.93	50.00	50.00	26.06
24	Dezhou	41.38	11.70	0.51	26.74	6.63	28.17	25.00	25.00	22.90
25	Pingxiang	39.76	11.24	0.69	36.17	2.50	10.61	25.00	25.00	20.76
26	Changzhi	35.83	10.13	0.51	26.60	4.54	19.29	25.00	25.00	20.25
27	Bengbu	20.46	5.78	0.22	11.40	6.67	28.35	25.00	25.00	17.63
28	Luohe	30.41	8.60	0.29	15.15	4.32	18.34	25.00	25.00	16.77
29	Huainan	11.79	3.33	0.12	6.42	5.63	23.94	25.00	25.00	14.67
30	Hebi	31.41	8.88	0.22	11.35	2.11	8.96	25.00	25.00	13.55
31	Wuhai	15.84	4.48	0.34	17.84	0.84	3.56	25.00	25.00	12.72
32	Liaoyuan	12.79	3.62	0.16	8.44	1.59	6.75	25.00	25.00	10.95
33	Liupanshui	26.34	7.45	0.49	25.38	2.55	10.85	0.00	0.00	10.92

Unit of a: %
Unit of b: none

Table A.22 Intelligent construction comprehensive evaluation and divided evaluation of 41 cities in the world

City	Overall		Intelligent environment and construction		Intelligent management and services		Intelligent economy and industry		Intelligent hardware facilities		Residents' intelligent potential	
	Ranking	Score	Ranking	Score	Ranking	Score	Ranking	Score	Ranking	Score	Ranking	Score
London	1	65.67	13	77.66	4	72.05	6	53.33	4	63.47	5	61.85
Amsterdam	2	65.51	1	97.84	3	72.86	5	54.14	10	56.56	27	46.13
Helsinki	3	64.01	10	84.84	1	74.98	8	47.55	23	47.53	3	65.15
Boston	4	63.87	6	88.42	2	74.36	3	59.63	31	41.29	14	55.65
Copenhagen	5	62.92	8	85.90	27	50.60	4	59.60	5	62.72	13	55.78
Vienna	6	61.22	4	92.03	15	68.98	22	35.83	7	60.97	25	48.30
Washington DC	7	60.92	24	67.79	19	61.54	1	75.58	24	45.85	23	53.84
Seattle	8	60.02	2	92.43	33	45.91	9	47.07	9	59.53	18	55.16
Chicago	9	59.04	18	75.70	13	70.09	17	42.38	17	51.82	17	55.19
San Jose	10	58.77	14	76.86	11	70.16	24	34.57	14	53.58	8	58.67
Portland	11	57.92	23	68.55	10	70.22	20	38.82	12	54.80	10	57.21
San Diego	12	57.09	17	76.00	5	71.60	28	33.11	21	49.69	19	55.06
Dubuque	13	56.62	26	62.50	8	70.95	34	27.43	18	50.19	1	72.04
Manchester	14	56.21	12	82.27	28	48.04	7	52.08	36	35.09	4	63.59
New York	15	55.51	22	72.10	32	45.95	14	44.12	8	60.83	22	54.56
Barcelona	16	55.22	20	75.00	26	53.71	10	46.70	2	68.73	28	31.98
Detroit	17	52.51	34	44.52	12	70.15	18	41.42	19	49.97	11	56.49
Minneapolis and Sao Paulo	18	52.16	27	62.50	20	59.85	15	43.67	35	39.16	15	55.61
Philadelphia	19	52.12	15	76.21	31	46.31	16	43.12	33	40.34	21	54.63
Ningbo	20	51.86	7	86.81	6	71.52	37	22.54	3	65.00	37	13.44
Issy-les-Moulineaux, Paris	21	51.39	38	25.00	23	58.19	30	30.95	1	71.27	2	71.55

(continued)

Table A.22 (continued)

City	Overall		Intelligent environment and construction		Intelligent management and services		Intelligent economy and industry		Intelligent hardware facilities		Residents' intelligent potential	
	Ranking	Score	Ranking	Score	Ranking	Score	Ranking	Score	Ranking	Score	Ranking	Score
San Francisco	22	50.96	25	66.71	30	47.12	13	45.27	32	40.89	20	54.83
Lisbon	23	49.51	28	62.50	16	67.05	29	31.90	6	62.43	34	23.69
Cleveland	24	48.46	35	37.50	21	59.08	19	40.52	20	49.84	16	55.37
Birmingham	25	47.48	37	28.14	7	71.20	12	45.73	40	32.51	6	59.79
Aarhus	26	47.40	29	62.50	38	25.07	11	45.79	26	45.25	9	58.40
Liverpool	27	46.65	40	20.02	22	58.41	2	61.21	37	34.10	7	59.53
Wuhan	28	46.23	11	82.48	14	69.19	35	27.06	30	41.86	41	10.58
Wuxi	29	45.85	9	85.18	17	62.89	23	35.31	38	33.31	38	12.58
Turin	30	45.33	31	50.00	18	61.84	27	33.97	13	53.60	32	27.26
Zhenjiang	31	45.21	5	89.75	29	47.34	25	34.51	29	42.51	39	11.95
Pudong, Shanghai	32	45.06	19	75.70	9	70.56	40	19.45	25	45.34	35	14.26
Jinhua	33	44.69	3	92.11	35	33.47	36	26.45	34	39.74	29	31.67
Taizhou	34	43.26	21	75.00	24	57.82	21	37.18	39	32.72	36	13.61
Kolner	35	43.16	32	50.00	40	21.93	26	34.18	15	53.39	12	56.32
Zhuhai	36	42.05	16	76.11	25	57.13	38	22.11	28	42.93	40	11.94
Lyon	37	41.70	30	53.09	36	32.01	33	29.22	22	48.01	26	46.19
Frederikshavn	38	36.38	39	25.00	34	34.04	32	29.28	27	43.26	24	50.30
Malaga	39	34.89	36	37.50	41	18.25	31	30.74	11	56.35	30	31.62
Santander	40	32.41	33	50.00	37	28.87	39	21.71	41	31.02	31	30.47
Verona	41	25.81	41	12.50	39	24.40	41	12.27	16	52.63	33	27.26

Table A.23 Intelligent environment and construction original data and score of 41 cities in the world

Ranking	City	City PM2.5/PM10 monitoring stations density		City grid management level of coverage		Residents' intelligent transportation tools usage level		Online publishing level of city future construction plan		Synthetical value
		Original data a	Score	Original data b	Score	Original data b	Score	Original data b	Score	
1	Amsterdam	0.0543	91.35	100.00	100.00	100.00	100.00	100.00	100.00	97.84
2	Seattle	0.0414	69.72	100.00	100.00	100.00	100.00	100.00	100.00	92.43
3	Jinhua	0.0407	68.45	100.00	100.00	100.00	100.00	100.00	100.00	92.11
4	Vienna	0.0405	68.11	100.00	100.00	100.00	100.00	100.00	100.00	92.03
5	Zhenjiang	0.0351	58.98	100.00	100.00	100.00	100.00	100.00	100.00	89.75
6	Boston	0.0319	53.67	100.00	100.00	100.00	100.00	100.00	100.00	88.42
7	Ningbo	0.0281	47.24	100.00	100.00	100.00	100.00	100.00	100.00	86.81
8	Copenhagen	0.0259	43.59	100.00	100.00	100.00	100.00	100.00	100.00	85.90
9	Wuxi	0.0242	40.74	100.00	100.00	100.00	100.00	100.00	100.00	85.18
10	Helsinki	0.0234	39.36	100.00	100.00	100.00	100.00	100.00	100.00	84.84
11	Wuhan	0.0178	29.90	100.00	100.00	100.00	100.00	100.00	100.00	82.48
12	Manchester	0.0173	29.10	100.00	100.00	100.00	100.00	100.00	100.00	82.27
13	London	0.0063	10.65	100.00	100.00	100.00	100.00	100.00	100.00	77.66
14	San Jose	0.0044	7.43	100.00	100.00	100.00	100.00	100.00	100.00	76.86
15	Philadelphia	0.0029	4.84	100.00	100.00	100.00	100.00	100.00	100.00	76.21
16	Zhuhai	0.0324	54.43	50.00	50.00	100.00	100.00	100.00	100.00	76.11
17	San Diego	0.0024	4.01	100.00	100.00	100.00	100.00	100.00	100.00	76.00
18	Chicago	0.0017	2.80	100.00	100.00	100.00	100.00	100.00	100.00	75.70
19	Pudong, Shanghai	0.0017	2.78	100.00	100.00	100.00	100.00	100.00	100.00	75.70
20	Barcelona	0.0000	0.00	100.00	100.00	100.00	100.00	100.00	100.00	75.00

(continued)

Table A.23 (continued)

Ranking	City	City PM2.5/PM10 monitoring stations density		City grid management level of coverage		Residents' intelligent transportation tools usage level		Online publishing level of city future construction plan		Synthetical value
		Original data a	Score	Original data b	Score	Original data b	Score	Original data b	Score	
21	Taizhou	0.0594	100.00	50.00	50.00	50.00	50.00	100.00	100.00	75.00
22	New York	0.0228	38.38	100.00	100.00	100.00	100.00	50.00	50.00	72.10
23	Portland	0.0144	24.18	100.00	100.00	100.00	100.00	50.00	50.00	68.55
24	Washington DC	0.0126	21.16	100.00	100.00	100.00	100.00	50.00	50.00	67.79
25	San Francisco	0.0100	16.83	100.00	100.00	100.00	100.00	50.00	50.00	66.71
26	Lisbon	0.0000	0.00	50.00	50.00	100.00	100.00	100.00	100.00	62.50
27	Minneapolis and Sao Paulo	0.0000	0.00	100.00	100.00	100.00	100.00	50.00	50.00	62.50
28	Dubuque	0.0000	0.00	50.00	50.00	100.00	100.00	100.00	100.00	62.50
29	Aarhus	0.0000	0.00	100.00	100.00	100.00	100.00	50.00	50.00	62.50
30	Lyon	0.0073	12.34	100.00	100.00	0.00	0.00	100.00	100.00	53.09
31	Turin	0.0000	0.00	0.00	0.00	100.00	100.00	100.00	100.00	50.00
32	Kolner	0.0000	0.00	0.00	0.00	100.00	100.00	100.00	100.00	50.00
33	Santander	0.0000	0.00	0.00	0.00	100.00	100.00	100.00	100.00	50.00
34	Detroit	0.0167	28.09	100.00	100.00	0.00	0.00	50.00	50.00	44.52
35	Cleveland	0.0000	0.00	0.00	0.00	100.00	100.00	50.00	50.00	37.50
36	Malaga	0.0000	0.00	0.00	0.00	100.00	100.00	50.00	50.00	37.50
37	Birmingham	0.0075	12.57	0.00	0.00	0.00	0.00	100.00	100.00	28.14
38	Issy-les-Moulineaux, Paris	0.0000	0.00	0.00	0.00	0.00	0.00	100.00	100.00	25.00
39	Frederikshavn	0.0000	0.00	0.00	0.00	0.00	0.00	100.00	100.00	25.00

(continued)

Table A.23 (continued)

Ranking	City	City PM2.5/PM10 monitoring stations density		City grid management level of coverage		Residents' intelligent transportation tools usage level		Online publishing level of city future construction plan		Synthetical value
		Original data a	Score	Original data b	Score	Original data b	Score	Original data b	Score	
40	Liverpool	0.0179	30.09	0.00	0.00	0.00	0.00	50.00	50.00	20.02
41	Verona	0.0000	0.00	0.00	0.00	0.00	0.00	50.00	50.00	12.50

Unit of a: pcs/km²
Unit of b: none

Table A.24 Intelligent management and service original data and score of 41 cities in the world

Ranking	City	Online publicity level of the government's non-classified documents		Online public participation ratio		Residents' health electronic archives usage level		Emergency intelligent level of emergency		Synthetical value
		Original data a	Score	Original data b	Score	Original data a	Score	Original data a	Score	
1	Helsinki	89.00	97.80	1.49	2.13	100.00	100.00	100.00	100.00	74.98
2	Boston	73.00	80.22	12.07	17.23	100.00	100.00	100.00	100.00	74.36
3	Amsterdam	83.00	91.21	0.17	0.24	100.00	100.00	100.00	100.00	72.86
4	London	76.00	83.52	3.28	4.68	100.00	100.00	100.00	100.00	72.05
5	San Diego	73.00	80.22	4.33	6.18	100.00	100.00	100.00	100.00	71.60
6	Ningbo	36.01	39.57	32.59	46.51	100.00	100.00	100.00	100.00	71.52
7	Birmingham	76.00	83.52	0.90	1.28	100.00	100.00	100.00	100.00	71.20
8	Dubuque	73.00	80.22	2.52	3.59	100.00	100.00	100.00	100.00	70.95
9	Pudong, Shanghai	71.51	78.58	2.57	3.67	100.00	100.00	100.00	100.00	70.56
10	Portland	73.00	80.22	0.47	0.67	100.00	100.00	100.00	100.00	70.22
11	San Jose	73.00	80.22	0.30	0.43	100.00	100.00	100.00	100.00	70.16
12	Detroit	73.00	80.22	0.26	0.37	100.00	100.00	100.00	100.00	70.15
13	Chicago	73.00	80.22	0.11	0.15	100.00	100.00	100.00	100.00	70.09
14	Wuhan	36.75	40.38	25.48	36.36	100.00	100.00	100.00	100.00	69.19
15	Vienna	69.00	75.82	0.06	0.09	100.00	100.00	100.00	100.00	68.98
16	Lisbon	62.00	68.13	0.05	0.06	100.00	100.00	100.00	100.00	67.05
17	Wuxi	34.62	38.04	44.49	63.50	50.00	50.00	100.00	100.00	62.89
18	Turin	43.00	47.25	0.08	0.11	100.00	100.00	100.00	100.00	61.84
19	Washington DC	73.00	80.22	11.17	15.95	100.00	100.00	50.00	50.00	61.54

(continued)

Table A.24 (continued)

Ranking	City	Online publicity level of the government's non-classified documents		Online public participation ratio		Residents' health electronic archives usage level		Emergency intelligent level of emergency		Synthetical value
		Original data a	Score	Original data b	Score	Original data a	Score	Original data a	Score	
20	Minneapolis and Sao Paulo	73.00	80.22	6.44	9.20	100.00	100.00	50.00	50.00	59.85
21	Cleveland	73.00	80.22	4.27	6.09	100.00	100.00	50.00	50.00	59.08
22	Liverpool	76.00	83.52	0.09	0.13	50.00	50.00	100.00	100.00	58.41
23	Issy-les-Moulineaux, Paris	71.00	78.02	3.31	4.73	100.00	100.00	50.00	50.00	58.19
24	Taizhou	28.46	31.27	70.06	100.00	50.00	50.00	50.00	50.00	57.82
25	Zhuhai	65.13	71.57	4.88	6.97	50.00	50.00	100.00	100.00	57.13
26	Barcelona	59.00	64.84	0.01	0.02	100.00	100.00	50.00	50.00	53.71
27	Copenhagen	91.00	100.00	1.68	2.40	100.00	100.00	0.00	0.00	50.60
28	Manchester	76.00	83.52	6.04	8.62	0.00	0.00	100.00	100.00	48.04
29	Zhenjiang	34.00	37.36	36.43	52.00	100.00	100.00	0.00	0.00	47.34
30	San Francisco	73.00	80.22	5.79	8.27	0.00	0.00	100.00	100.00	47.12
31	Philadelphia	73.00	80.22	3.53	5.03	100.00	100.00	0.00	0.00	46.31
32	New York	73.00	80.22	2.51	3.59	50.00	50.00	50.00	50.00	45.95
33	Seattle	73.00	80.22	2.39	3.41	0.00	0.00	100.00	100.00	45.91
34	Frederikshavn	78.00	85.71	0.30	0.43	50.00	50.00	0.00	0.00	34.04
35	Jinhua	28.10	30.88	2.12	3.02	100.00	100.00	0.00	0.00	33.47
36	Lyon	71.00	78.02	0.03	0.04	50.00	50.00	0.00	0.00	32.01

(continued)

Table A.24 (continued)

Ranking	City	Online publicity level of the government's non-classified documents		Online public participation ratio		Residents' health electronic archives usage level		Emergency intelligent level of emergency		Synthetical value
		Original data a	Score	Original data b	Score	Original data a	Score	Original data a	Score	
37	Santander	59.00	64.84	0.44	0.63	50.00	50.00	0.00	0.00	28.87
38	Aarhus	91.00	100.00	0.19	0.27	0.00	0.00	0.00	0.00	25.07
39	Verona	43.00	47.25	0.24	0.34	50.00	50.00	0.00	0.00	24.40
40	Kolner	78.00	85.71	1.41	2.01	0.00	0.00	0.00	0.00	21.93
41	Malaga	59.00	64.84	5.72	8.16	0.00	0.00	0.00	0.00	18.25

Unit of a: none
Unit of b: %

Table A.25 Intelligent economy and industry original data and score of 41 cities in the world

Ranking	City	The proportion of R&D expenditure in GDP		City labor productivity		City product value density		The proportion of city intelligent industry		Synthetical value
		Original data a	Score	Original data b	Score	c Original data	Score	Original data a	Score	
1	Washington DC	2.79	78.59	642278	100.00	2345762712	100.00	1.39	23.72	75.58
2	Liverpool	1.72	48.45	289058	45.01	1205481044	51.39	5.86	100.00	61.21
3	Boston	2.79	78.59	496466	77.30	1381732012	58.90	1.39	23.72	59.63
4	Copenhagen	3.10	87.32	204545	31.85	1351508121	57.61	3.61	61.60	59.60
5	Amsterdam	2.16	60.85	396159	61.68	1469542221	62.65	1.84	31.40	54.14
6	London	1.72	48.45	86877	13.53	1204712085	51.36	5.86	100.00	53.33
7	Manchester	1.72	48.45	175582	27.34	763510592	32.55	5.86	100.00	52.08
8	Helsinki	3.55	100.00	122949	19.14	105522090	4.50	3.90	66.55	47.55
9	Seattle	2.79	78.59	380508	59.24	627302275	26.74	1.39	23.72	47.07
10	Barcelona	3.00	84.51	105494	16.42	1678115800	71.54	0.84	14.33	46.70
11	Aarhus	3.10	87.32	123455	19.22	352395604	15.02	3.61	61.60	45.79
12	Birmingham	1.72	48.45	104639	16.29	426858871	18.20	5.86	100.00	45.73
13	San Francisco	2.79	78.59	366115	57.00	510489510	21.76	1.39	23.72	45.27
14	New York	2.79	78.59	145141	22.60	1210000000	51.58	1.39	23.72	44.12
15	Minneapolis and Sao Paulo	2.79	78.59	287578	44.77	646878474	27.58	1.39	23.72	43.67
16	Philadelphia	2.79	78.59	208735	32.50	883378747	37.66	1.39	23.72	43.12
17	Chicago	2.79	78.59	194614	30.30	865676568	36.90	1.39	23.72	42.38
18	Detroit	2.79	78.59	270596	42.13	498067724	21.23	1.39	23.72	41.42
19	Cleveland	2.79	78.59	254527	39.63	472846442	20.16	1.39	23.72	40.52

(continued)

Table A.25 (continued)

Ranking	City	The proportion of R&D expenditure in GDP		City labor productivity		City product value density		The proportion of city intelligent industry		Synthetical value
		Original data a	Score	Original data b	Score	c Original data	Score	Original data a	Score	
20	Portland	2.79	78.59	238619	37.15	370695620	15.80	1.39	23.72	38.82
21	Taizhou	1.98	55.77	52760	8.21	484485337	20.65	3.75	64.08	37.18
22	Vienna	2.39	67.32	101832	15.85	433618715	18.49	2.44	41.64	35.83
23	Wuxi	1.98	55.77	48427	7.54	324380843	13.83	3.75	64.08	35.31
24	San Jose	2.79	78.59	145422	22.64	312158744	13.31	1.39	23.72	34.57
25	Zhenjiang	1.98	55.77	40371	6.29	279199628	11.90	3.75	64.08	34.51
26	Kolner	2.92	82.25	105689	16.46	267222017	11.39	1.56	26.62	34.18
27	Turin	3.00	84.51	84117	13.10	589229469	25.12	0.77	13.14	33.97
28	San Diego	2.79	78.59	139981	21.79	195332345	8.33	1.39	23.72	33.11
29	Lisbon	1.50	42.25	173840	27.07	1122641509	47.86	0.61	10.41	31.90
30	Issy-les-Moulineaux, Paris	2.26	63.66	64263	10.01	235236220	10.03	2.35	40.10	30.95
31	Malaga	1.30	36.62	331736	51.65	477453165	20.35	0.84	14.33	30.74
32	Frederikshavn	2.92	82.25	42978	6.69	36446860	1.55	1.56	26.62	29.28
33	Lyon	2.26	63.66	58276	9.07	94740041	4.04	2.35	40.10	29.22
34	Dubuque	2.79	78.59	39905	6.21	28444225	1.21	1.39	23.72	27.43
35	Wuhan	1.98	55.77	16192	2.52	177364336	7.56	2.48	42.40	27.06
36	Jinhua	1.98	55.77	73493	11.44	462647616	19.72	1.11	18.87	26.45
37	Ningbo	1.98	55.77	31373	4.88	247463711	10.55	1.11	18.94	22.54
38	Zhuhai	1.98	55.77	26175	4.08	158335063	6.75	1.28	21.84	22.11

(continued)

Table A.25 (continued)

Ranking	City	The proportion of R&D expenditure in GDP		City labor productivity		City product value density		The proportion of city intelligent industry		Synthetical value
		Original data a	Score	Original data b	Score	c Original data	Score	Original data a	Score	
39	Santander	1.30	36.62	96204	14.98	490542857	20.91	0.84	14.33	21.71
40	Pudong, Shanghai	1.98	55.77	17259	2.69	73789841	3.15	0.95	16.21	19.45
41	Verona	1.27	35.77	754	0.12	967914	0.04	0.77	13.14	12.27

Unit of a: %
Unit of b: dollars/person
Unit of c: dollars/km^2

Table A.26 Intelligent hardware facilities original data and score of 41 cities in the world

Ranking	City	Public space free network coverage density		Mobile network per capita usage		City broadband speed		Intelligent grid level of coverage		Synthetical value
		Original data a	Score	Original data b	Score	Original data c	Score	Original data d	Score	
1	Issy-les-Moulineaux, Paris	1.1765	17.78	82.77	67.31	104.81	100.00	100.00	100.00	71.27
2	Barcelona	2.6104	39.45	113.98	92.69	44.83	42.77	100.00	100.00	68.73
3	Ningbo	6.6175	100.00	41.15	33.46	27.83	26.55	100.00	100.00	65.00
4	London	2.2209	33.56	114.3	92.95	28.70	27.38	100.00	100.00	63.47
5	Copenhagen	1.1369	17.18	106.5	86.61	49.37	47.10	100.00	100.00	62.72
6	Lisbon	1.1439	17.29	114.51	93.12	41.20	39.31	100.00	100.00	62.43
7	Vienna	0.6753	10.20	112.97	91.87	43.82	41.81	100.00	100.00	60.97
8	New York	1.4560	22.00	83.93	68.25	55.61	53.06	100.00	100.00	60.83
9	Seattle	0.1426	2.15	108.11	87.92	52.44	50.03	100.00	100.00	60.03
10	Amsterdam	1.4410	21.77	83.93	68.25	50.40	48.09	100.00	100.00	59.53
11	Malaga	0.4058	6.13	95.13	77.36	44.80	42.74	100.00	100.00	56.56
12	Portland	0.0127	0.19	113.98	92.69	34.09	32.53	100.00	100.00	56.35
13	Turin	1.0671	16.13	83.93	68.25	36.48	34.81	100.00	100.00	54.80
14	San Jose	0.1460	2.21	122.97	100.00	12.78	12.19	100.00	100.00	53.60
15	Kolner	0.4527	6.84	83.93	68.25	41.10	39.21	100.00	100.00	53.58
16	Verona	0.1506	2.28	102.34	83.22	29.40	28.05	100.00	100.00	53.39
17	Chicago	0.0194	0.29	122.97	100.00	10.70	10.21	100.00	100.00	52.63
18	Dubuque	0.6766	10.22	83.93	68.25	30.20	28.81	100.00	100.00	51.82
19	Detroit	0.0866	1.31	83.93	68.25	32.70	31.20	100.00	100.00	50.19

(continued)

Table A.26 (continued)

Ranking	City	Public space free network coverage density		Mobile network per capita usage		City broadband speed		Intelligent grid level of coverage		Synthetical value
		Original data a	Score	Original data b	Score	Original data c	Score	Original data d	Score	
20	Cleveland	0.4567	6.90	83.93	68.25	25.90	24.71	100.00	100.00	49.97
21	San Diego	0.6367	9.62	83.93	68.25	22.50	21.47	100.00	100.00	49.84
22	Lyon	0.2260	3.42	83.93	68.25	28.40	27.10	100.00	100.00	49.69
23	Helsinki	0.1100	1.66	82.77	67.31	76.57	73.06	50.00	50.00	48.01
24	Washington DC	2.0000	30.22	83.93	68.25	36.60	34.92	50.00	50.00	45.85
25	Pudong, Shanghai	1.0733	16.22	41.15	33.46	35.19	33.58	100.00	100.00	45.81
26	Aarhus	0.8264	12.49	41.15	33.46	37.10	35.40	100.00	100.00	45.34
27	Frederikshavn	0.1648	2.49	106.5	86.61	43.93	41.91	50.00	50.00	45.25
28	Zhuhai	0.2146	3.24	102.34	83.22	38.35	36.59	50.00	50.00	43.26
29	Zhenjiang	0.6319	9.55	41.15	33.46	30.10	28.72	100.00	100.00	42.93
30	Wuhan	1.0280	15.53	41.15	33.46	22.07	21.06	100.00	100.00	42.51
31	Boston	0.6015	9.09	41.15	33.46	26.07	24.87	100.00	100.00	41.86
32	San Francisco	0.9134	13.80	83.93	68.25	34.70	33.11	50.00	50.00	41.29
33	Philadelphia	0.6127	9.26	83.93	68.25	37.80	36.07	50.00	50.00	40.89
34	Jinhua	0.4850	7.33	83.93	68.25	37.50	35.78	50.00	50.00	40.34
35	Minneapolis and Sao Paulo	3.1214	47.17	41.15	33.46	29.67	28.31	50.00	50.00	39.74
36	Manchester	0.9872	14.92	83.93	68.25	24.60	23.47	50.00	50.00	39.16
37	Liverpool	1.4008	21.17	114.3	92.95	27.50	26.24	0.00	0.00	35.09
38	Wuxi	0.8852	13.38	114.3	92.95	31.50	30.05	0.00	0.00	34.10

(continued)

Table A.26 (continued)

Ranking	City	Public space free network coverage density		Mobile network per capita usage		City broadband speed		Intelligent grid level of coverage		Synthetical value
		Original data a	Score	Original data b	Score	Original data c	Score	Original data d	Score	
39	Taizhou	1.2323	18.62	41.15	33.46	30.16	28.78	50.00	50.00	32.72
40	Birmingham	0.3697	5.59	114.3	92.95	33.00	31.49	0.00	0.00	32.51
41	Santander	0.4286	6.48	113.98	92.69	26.13	24.93	0.00	0.00	31.02

Unit of a: pcs/km^2
Unit of b: %
Unit of c: Mbps
Unit of d: none

Table A.27 Residents' intelligent potential original data and score of 41 cities in the world

Ranking	City	The proportion of Internet users in city		The proportion of information professionals		The proportion of junior college or above educational level population		Residents' per capita online shopping expenditure amount		Synthetical value
		Original data a	Score	Original data a	Score	Original data a	Score	Original data b	Score	
1	Dubuque	77.50	47.45	41.60	75.23	89.30	100.00	586.12	65.47	72.04
2	Issy-les-Moulineaux, Paris	90.00	55.10	55.29	100.00	71.60	80.18	455.91	50.92	71.55
3	Helsinki	91.50	56.02	6.00	10.85	83.70	93.73	895.29	100.00	65.15
4	Manchester	89.80	54.98	9.94	17.98	76.80	86.00	853.93	95.38	63.59
5	London	89.80	54.98	6.10	11.03	76.80	86.00	853.93	95.38	61.85
6	Birmingham	89.80	54.98	1.56	2.81	76.80	86.00	853.93	95.38	59.79
7	Liverpool	89.80	54.98	0.96	1.74	76.80	86.00	853.93	95.38	59.53
8	San Jose	79.70	48.80	11.30	20.44	89.30	100.00	586.12	65.47	58.67
9	Aarhus	94.60	57.92	7.70	13.92	76.90	86.11	677.33	75.65	58.40
10	Portland	86.10	52.72	5.90	10.67	89.30	100.00	586.12	65.47	57.21
11	Detroit	78.40	48.00	6.90	12.48	89.30	100.00	586.12	65.47	56.49
12	Kolner	84.00	51.43	18.20	32.91	86.30	96.64	396.71	44.31	56.32
13	Copenhagen	95.00	58.16	1.76	3.18	76.90	86.11	677.33	75.65	55.78
14	Boston	86.20	52.78	2.40	4.34	89.30	100.00	586.12	65.47	55.65
15	Minneapolis and Sao Paulo	82.10	50.27	3.70	6.69	89.30	100.00	586.12	65.47	55.61
16	Cleveland	76.70	46.96	5.00	9.04	89.30	100.00	586.12	65.47	55.37
17	Chicago	78.50	48.06	4.00	7.23	89.30	100.00	586.12	65.47	55.19
18	Seattle	85.70	52.47	1.50	2.71	89.30	100.00	586.12	65.47	55.16

(continued)

Table A.27 (continued)

Ranking	City	The proportion of Internet users in city		The proportion of information professionals		The proportion of junior college or above educational level population		Residents' per capita online shopping expenditure amount		Synthetical value
		Original data a	Score	Original data a	Score	Original data a	Score	Original data b	Score	
19	San Diego	79.70	48.80	3.30	5.97	89.30	100.00	586.12	65.47	55.06
20	San Francisco	79.70	48.80	2.80	5.06	89.30	100.00	586.12	65.47	54.83
21	Philadelphia	77.80	47.63	3.00	5.43	89.30	100.00	586.12	65.47	54.63
22	New York	81.50	49.90	1.60	2.89	89.30	100.00	586.12	65.47	54.56
23	Washington DC	76.50	46.84	170	3.07	89.30	100.00	586.12	65.47	53.84
24	Frederikshavn	98.40	60.25	0.00	0.00	86.30	96.64	396.71	44.31	50.30
25	Vienna	80.60	49.35	3.96	7.17	82.50	92.39	396.71	44.31	48.30
26	Lyon	81.90	50.14	1.93	3.50	71.60	80.18	455.91	50.92	46.19
27	Amsterdam	94.00	57.55	4.36	7.89	72.30	80.96	341.23	38.11	46.13
28	Barcelona	75.00	45.92	2.19	3.96	54.00	60.47	157.25	17.56	31.98
29	Jinhua	163.33	100.00	1.27	2.30	16.25	18.20	55.35	6.18	31.67
30	Malaga	71.60	43.84	2.55	4.61	54.00	60.47	157.25	17.56	31.62
31	Santander	71.60	43.84	0.00	0.00	54.00	60.47	157.25	17.56	30.47
32	Turin	58.50	35.82	0.00	0.00	56.00	62.71	94.19	10.52	27.26
33	Verona	58.50	35.82	0.00	0.00	56.00	62.71	94.19	10.52	27.26
34	Lisbon	62.10	38.02	0.00	0.00	35.00	39.19	157.25	17.56	23.69
35	Pudong, Shanghai	78.25	47.91	0.28	0.51	2.18	2.44	55.35	6.18	14.26
36	Taizhou	63.30	38.76	0.86	1.56	7.09	7.94	55.35	6.18	13.61
37	Ningbo	65.55	40.13	0.40	0.72	6.01	6.73	55.35	6.18	13.44

(continued)

Table A.27 (continued)

Ranking	City	The proportion of Internet users in city		The proportion of information professionals		The proportion of junior college or above educational level population		Residents' per capita online shopping expenditure amount		Synthetical value
		Original data a	Score	Original data a	Score	Original data a	Score	Original data b	Score	
38	Wuxi	60.43	37.00	0.34	0.61	5.82	6.52	55.35	6.18	12.58
39	Zhenjiang	46.99	28.77	0.30	0.54	10.98	12.30	55.35	6.18	11.95
40	Zhuhai	39.84	24.39	1.10	1.99	13.58	15.21	55.35	6.18	11.94
41	Wuhan	29.92	18.32	0.40	0.72	15.26	17.09	55.35	6.18	10.58

Unit of a: %
Unit of b: dollars/person

A6 An Overview and Ranking of Intelligent City Construction in China and in the World

Please visit http://www.Intelligent City Evaluation.org to see the details of a review and evaluation score of intelligent construction in this book (see Fig. A.3 for QR code). The website will annually release world's intelligent city ranking list.

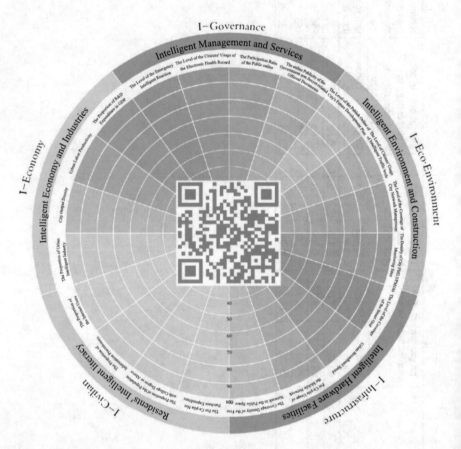

Fig. A.3 QR code of intelligent city website

A7 R&D Case of Intelligent City Evaluation Indicator System—Pudong, Shanghai[18]

As a new mode and a new path of city development, intelligent city construction has been attracting the attention of domestic cities and being sought after in recent years. More and more cities take intelligent city construction as a strategic choice of local development of social economic transition. With the rapid development and application of information technology and under the trend of national policy, Pudong New Area has put forward a preliminary conception of constructing "Intelligent City" in 2009. In 2011, Pudong New Area took the lead putting forward *Intelligent Pudong Construction Outline (iPudong2015), Carry Forward Intelligent Pudong Construction 2011–2013 Action Plan*, which is focusing on the top-level design of appropriately advanced intelligent city construction.

To organize the contents of intelligent city construction better and measure the development level of intelligent city, China has released some indicator systems for intelligent city successively since 2011. The setting of standard is just one of the practices and results of putting emphasis on intelligent government affairs. The basic and final purpose of exploring a more scientific intelligent government affairs development pattern, which is more suitable for national conditions, is promoting the scientific development of intelligent city. This built the basis for setting up of the standards and paved the way for standardization during the practices of promoting intelligent city construction.

In 2011, Shanghai Pudong Intelligent City Development Research Institute took the lead to release *Intelligent City Evaluation Indicator System*, which was improved in 2012. Dozens of indicator data from *Intelligent City Evaluation Indicator System 1.0* released in 2011 and *Intelligent City Evaluation Indicator System 2.0* released in 2012 provided references for intelligent city construction.

From *Intelligent City Evaluation Indicator System 1.0* to *Intelligent City Evaluation Indicator System 2.0*, the research continued for about one year. Version 2.0 added further improvements and testing evaluations on the basis of version 1.0, and put forward six dimensions for intelligent city evaluation: intelligent city infrastructures, intelligent city public administration and services, intelligent city information service economy development, intelligent city humane studies accomplishment, intelligent city residents' subjective perception, intelligent city soft environment construction. The modifications of the six dimensions and relevant segmented indicators helped version 2.0 "feel the pulse" of intelligent city construction process in China deeper and more precisely.

---

[18]The materials in this section are based on modification and improvement of Six Dimensions of Pudong, "Feel the Pulse" of Intelligent City (Authors: Sheng Xuefeng, Yang Xinmin). The original is available on China Informatization, 2012 No. 14.

A7.1 Trinity—Government, Society and Residents

Intelligent city evaluation indicator system is the standard to comprehensively reflect and measure intelligent city construction development stage and level, which has measuring and guiding function for intelligent city construction. Therefore, when designing indicator system frame and selecting specific indicators, not only their guidance for the government promoting intelligent city construction should be considered, they should also be fully suitable for various aspects of operation and experience for intelligent city construction, so the typical representative indicators should be selected form three aspects: government promotion, social participation, public perception, which should not only focusing on collectability and comparability of indicators, but also the history and current data should be collected reliably, conveniently and scientifically and be comparable within different cities and districts.

Hence, three aspects (government, society and citizens) were centered on during the further improvement of intelligent city evaluation indicator system, which put forward a three-level indicator system with six dimensions as the core.

The frame system covers the subjects of each level such as construction, operation, management and perception of intelligent city, and limits the total number of indicators strictly trying to reflect most current situations with least indicators. When designing indicators, compare quantitative indicators with qualitative indicators, and focus on complementation and reflection between specific indicators, try to reduce systematic errors caused by individual indicators.

A7.2 Cling to Construction Status, Lead the Future Development

The six dimensions of intelligent city evaluation indicator system has fully considered intelligent city infrastructures, demonstration application, industrial development and public perception in aspect of coverage, and reflected considering the construction status and leading the future development when selecting indicators and setting reference values.

1. Intelligent City Infrastructures

In the aspect of broad sense, intelligent city infrastructures refer to related infrastructures that ensure each function of intelligent city working together smoothly and safely. So to speak, all the infrastructures playing a role in intelligent city are included. While looking at the current status of intelligent city construction in China, the intelligent city infrastructures that we are now focusing on mainly include construction and application level of various wired and wireless broadband network.

Optical fiber broadband and wireless broadband are the core and basis of intelligent city (or digital city, smart city) construction in China. Many cities, including Shanghai, have regarded "optical network city" and "wireless city" as the basic function and safeguard of intelligent city construction. To scientifically reflect the city's basic network construction and application level, it put forward "family optical fiber access rate", "WLAN coverage rate of major public places" and "network access level of per family" from two aspects—the level of broadband network coverage and access level, and abandoned "average wireless network access broadband", "the proportion of network infrastructures investment in total investment of social fixed assets" and other indicators that were relatively repetitive and hard to collected. At same time, it should also refer to the city network construction and application situations domestic and overseas to put forward reference values for intelligent city construction. For example, both "family optical fiber access rate" and "WLAN coverage rate of major public places" should be higher than 99%, i.e. basically realizing full coverage, "network access level of per family" should be 30 M or above, which are also development goals in several future years of many cities in China.

2. Intelligent City Public Administration and Services

Intelligent city public administration and services are the core areas of the intelligent city construction. They involves many aspects, such as intelligent government administration, road traffic, health care, education, environmental monitoring, safety monitoring and controlling, energy management, social insurance and etc., which directly influence city residents' happiness and city management operation efficiency. At present, the mode of intelligent city construction planning in China is segmenting intelligent city public administration and services into several professional fields. For example, "intelligent Pudong" is planning to promote the construction of nine demonstration application engineering such as "Governmental Service Collaboration Engineering", while Ningbo has put forward ten application systems construction including "build intelligent logistics system" in intelligent city decisions. Under this frame system, we also fully combined with the intelligent city construction mode in China, and considered eight aspects individually—intelligent governmental services, traffic management, medical system, environmental protection, energy management, city safety, education system and community management, according to the basic frame of city intelligent application, and selected several specific indicators that were representative and that could fully reflect the construction application situation.

On the one hand, the designing of indicators mainly reflects the combination with current intelligent city construction concepts and planning in China, and managed to target at construction contents that are being built by intelligent cities in China or that are considered as important matters would be promoted in the future. For example, for intelligent governmental services, "online transaction proportion of administrative examination and approval matters" and "online flow rate of the government's non-classified documents" are selected, with the reference values set to 90 and 100% or above respectively, for intelligent traffic management, "electronic rate of bus stop board" and "citizens' compliance rate for traffic routing information" are selected, with the reference values set to 80 and 50% or above respectively, to provide references for intelligent city construction and application level. On the other hand, some perspective and leading indicators are designed in view of the intelligent city construction and development trend. For example, for intelligent energy management, "new energy automobile proportion" and "building digital energy saving proportion" were included in the indicator system, and they were set to the reference values of 10 and 30% or above. For intelligent education system, "network teaching proportion" was put forward and the reference value of it was set to 50% or above. We will guide the domestic intelligent city construction to pay more attention to these aspects through these perspective and leading indicators.

3. Intelligent City Information Service Economy Development

Intelligent city information service economy development presents a relationship of mutual promotion and interdependence to some extent. On the one hand, intelligent city construction depends on research and application of new technologies and new products; on the other hand, intelligent city construction operation will greatly promote the development of these industries, especially the development of information industry.

Therefore, we think the development of related industrial economies in intelligent city is an important factor to measure intelligent city construction level. Given that the scope of industrial economies related to intelligent city is wide, which is involved in electronic and information manufacturing, software information service and various aspects, and that there are huge differences among industrial structures of cities, if all the industries are included, there will be less comparability between cities. Therefore, when designing the indicator system, we should mainly consider the development situation of information service industry that derived from intelligent city construction and development or that supports intelligent city construction operation, which mainly includes the following two aspects.

(1) The overall level of industrial development, which refers to overall strength of the development of the city information services. Specific considerations indicators include "the proportion of added value of information services industry in gross regional production" and "the proportion of information service industry employees in total social employees", and the reference values of both are 10% or above. The overall level of information service development can be reflected by these two indicators.

(2) Enterprise informatization operating level, which refers to the development level that supports enterprise's production and operation through informatization system. It mainly includes three specific indicators—"the building rate of enterprise website", "enterprise electronic commerce behavior rate" and "enterprise informatization system usage rate", and the reference values of them are set to 90, 95 and 90% or above respectively. At the current stage of

intelligent city construction in China, the above three indicators can reflect the enterprise informatization operation level better.

4. Intelligent City Humane Studies Accomplishment

Intelligent city humane studies accomplishment is mainly used to measure citizens' cognition of intelligent city development concepts, the mastery of basic science and technology (including information technology) as well as intelligent life concepts. As the main body of intelligent city operation and service, citizens' own situation is decisive for the successful construction of intelligent city. At the same time, the intelligence of their behavior is most important factors that directly reflect intelligent city construction results. To this end, we will talk about intelligent city humane studies accomplishment level.

(1) Residents income level. Although resident's income level is the indicator of city economic development, we think residents' income will have enormous influence on city management and life. It's hard to think intelligent city management mode and resident's life style can be built up in a city where the resident's income level is very low. According to resident's disposable income level in Shanghai and other regions, the reference value of intelligent city residents income is about 50 thousand Yuan or above.
(2) Residents culture science literacy. Resident's culture science literacy includes knowledge of natural sciences, social sciences and other aspects. It is an indicator that can comprehensively reflect resident's culture science literacy, which plays an important role in intelligent city construction. "The proportion of specific colleague course or above in gross population" is selected as the indicator to reflect residents culture science literacy. Combined with the situations of Shanghai and other major cities, the reference value of this indicator is set to 30% or above.
(3) Residents life networked level. Networked life is an important feature of intelligent city. Therefore, When investigating the intelligent city citizen culture science literacy, resident's life networked level is an important reference system; especially "resident's network access rate" and "family online-shopping proportion", they should become characteristic indicators of resident's life networked level. So, the reference values of both are set to 60% or above.

5. Intelligent City Residents' Subjective Perception

Intelligent city residents' subjective perception gives priority to the indicator of resident's subjective perception. Evaluating and measuring important aspects related to intelligent city construction are critical reflections of residents' happiness. Indicators related to intelligent city residents' subjective perception are important complements to other indicators that are not related to subjective perception. They can reflect intelligent city construction results more precisely and more intuitively, and they are important ways to reflect the spirit of "people oriented intelligent city". Indicators shall be designed from both sense of convenience and security, using the mode of "sample investigation + subjective rating" to acquire the results of intelligent city construction in the mind of residents.

(1) The sense of convenience of life mainly refers to convenience degree in various aspects such as traveling, seeking medical treatment and handling affairs. Indicators shall be designed according to current main focusing points of city development—transportation, medical and government services, and let residents rate the convenience degree of obtaining traffic information, medical treatment and governmental services, and the reference value is set to 8 points or above (the total points are 10).

(2) The sense of security of life mainly refers to the degree of satisfaction of residents to intelligence level of food and drug safety, environmental safety, transportation safety and etc. In recently years, food and drug safety, environmental safety and transportation safety are three major fields in city management operation and public life. One of the important goals of intelligent city is to protect these three safeties through intelligent application system.

Therefore, modified indicators include "food and drug safety electronic monitoring satisfaction degree", "environmental safety information monitoring satisfaction degree" and "traffic safety information system satisfaction degree", which reflect intelligent city management application system construction and operation level from the perspective of subjective perception of residents.

6. Intelligent City Soft Environment Construction

Intelligent city soft environment construction mainly consists of planning, design and environmental building of intelligent city development. Now, China is in an early stage of intelligent city construction, the status of overall planning and design and soft environment construction such as environmental building will have important influence on intelligent city construction. Therefore, modified indicator system has contained three specific indicators—"intelligent city development planning", "intelligent city organizational leadership mechanism" and "intelligent city forum conference and training level", in the view of intelligent city planning and design and intelligent city atmosphere building. These three specific indicators could reflect the soft environment power of the city.

A7.3 The Indicator System Needs to Be Modified and Improved Continuously Through Empirical Research

Intelligent city construction is not achieved overnight. It needs long-time investment and construction as well as focusing promotion. Therefore, when performing intelligent city evaluation, considering some characteristics such as collectability and comparability, indicators are classified. Some indicators of greater importance are characterized as "core indicators", and others were characterized as "general indicators", and they will be given different weights during evaluation. For evaluation results, we pay more attention to the current stage of development, and divide the evaluation results into three kinds—incubation period, hatching period and embryonic period.

Through collecting and test-evaluating the indicators from Shanghai Pudong New Area, Hangzhou and other regions, the indicator system more truly reflects the intelligent city construction stage and level, and finds out weaknesses in current intelligent city construction to some extent. For example, the subjective perception of residents in current intelligent city construction is still relatively weak. This means intelligent city construction is still staying in the government level to a large extent, both promotion and effectiveness have not yet deeply rooted in the hearts of people. The scores related to intelligent energy management are generally low, which shows that efficiency of energy conservation and emissions reduction or smart grid construction needs to be accelerated. Empirical test evaluation is the inspection of the indicator system, which means some measuring and guiding significance to intelligent city construction. Meanwhile, empirical research also means a lot to the further improvement of the indicator system.

Intelligent city construction is a long and complicated process. As an important tool for measuring and guiding intelligent city construction, the indicator system also needs to be updated and changed continuously. Not only the system frame should be more completed and scientific, but also the indicator selection should be more typical and efficient.

Only when we improve the intelligent city evaluation indicators system during the development and perfect it constantly during practices, it could truly keep pace with the times and measure the intelligent city construction stage and level for playing a leading role in city innovation and development in China.

A8   R&D Case of Intelligent City Evaluation Indicator System—TU Wien[19]

The intelligent city evaluation indicators system—Smart City Indicators, is led and developed by Regional Research Center in Vienna University of Technology (TU Wien) through cooperation with multiple scientific research departments and local cities. Since the first version of the indicator system and evaluation report was published in 2007, two subsequent updated versions have been released in 2013 and 2014 respectively. This evaluation indicator system is research result of the university scientific research team, which is funded by public and private funds. Most of the research results and part of the processes were published to the public. At the same time, their research results were also promoted and applied in the researches and practices of other intelligent cities.

Since 2007, the team led by Prof. Rudolf Giffinger, the director of Regional Research Center in Vienna University of Technology, has started to carry out related researches and built up an intelligent city model under the background of European city development based on their understanding of intelligent city. Considering the large proportion of medium-sized cities in European city residents, as well as the feasibility of the research and data availability, the research focused on sustainable evaluation and ranking of medium-sized European cities. The main evaluation and ranking results were published and updated on the official website of European intelligent cities established by the team.[20] As a scholar who has geography and city planning background, Prof. Giffinger understood and built the intelligent city model from the perspective of city planning. When the first version of this evaluation system was published in 2007, it was based on model design and divided into three hierarchies. Based on the premise of keeping the basic frame, the second version published in 2013 has been improved a lot. The understanding of three hierarchies was updated, and these changes were kept in the third version published in 2014. Prof. Giffinger thought the model and evaluation system was a good way to measure and research the innovative performance in all respects of objective cities.

A8.1 Research Process

At first, evaluating and ranking an intelligent city should be based on the understanding of the concept of intelligent city. Economy and technology development under the background of globalization have profoundly influenced the European cities, presenting double challenges to European cities—city competitiveness

---

[19]The materials in this section are based on the interview with Prof. RudolfGiffinger from Vienna University of Technology (Author: Lü Hui). Thanks.
[20]Please go to http://www.smart-cities.eu for details.

improvement and sustainable development. While these challenges are closely bound up with all respects of cities, especially problems related with life quality of residents, such as housing, economy, culture, society, environment and etc. The concept of intelligent city is originated from the development of information and communication technologies. Cities are looking forward to dealing with multiple challenges they are facing through technologies. Prof. Giffinger thinks technology is just one dimension of intelligent city, while social innovation dominated by self-organizing and learning process is an indispensable part of intelligent city. We need to understand intelligent city through a more comprehensive perspective. The reason for choosing to rank cities—a seemingly non-academic method, is that in the combination of city research and planning practice, ranking is possible to become a catalyst. Meanwhile, Prof. Giffinger hopes to make it an effective tool through city evaluation and ranking, promoting horizontal comparison between cities in the intelligent city model combing with development status of each city to find development direction and covert the evaluation ranking into a reference system for specific development strategies of cities.

1. The First Version of Evaluation Indicator System

Vienna University of Technology has published a research report—*Smart Cities: Ranking of European Medium-Sized Cities* in cooperation with its research partners in October, 2007. The research participants included University of Ljubljana in Slovenia and Delft University of Technology in Netherlands.

The aspects and indicators of intelligent city evaluation used in the research are closely related to the target cities. Because there were no evaluation and ranking for intelligent city in the beginning of the research, the evaluation and ranking system of 7 cities that have more influences than considered first, and they selected evaluation ranking results of years near the research year to investigate, see Table A.28.

In Table A.28, some (such as 1, 3, 4, 6) mainly focus on life quality of individual residents in the city, while others (such as 5 and 7) include a wider range of factors, such as geographic elements and tourist attraction and etc., and the 2nd evaluation ranking focuses on a particular aspect of the city—the sustainability of the city environment.

The evaluation ranking limits the evaluation scope to a certain spatial scale, such as global scale or a country scale. Because it's difficult to evaluate and compare all the cities through one method, generally, cities would be classified according to their population size (such as 2, 5, 7), or they will select target cities to be evaluated according to their importance (such as 1, 3, 4). The selection method of 6th ranking is more comprehensive, perform preliminary evaluation for 643 cities in Europe first, then select 58 cities to perform actual evaluation and ranking according to the results of preliminary evaluation. Data availability is another factor that will impact evaluation method. Some evaluations (such as 1 and 3) get data through field survey and the interview, while most of them (such as 2, 4, 5, 6, 7) use data analysis research. For evaluations, to determine the weight of each factor is an important aspect. The weights in most evaluations are determined by the research team, while

Table A.28 Evaluation ranking results referenced in the research

No.	Title	Author	Year published	Scope
1	Quality of living survey	Mercer Human Resource Consulting	2007	200 cities in the world
2	Canada's most sustainable cities	Corporate Knights: The Canadian Magazine for Responsible Business	2007	Large city centers in Canada
3	How the world views its cities	Anholt City Brands	2006	60 cities in the world
4	Worldwide cost of living	Economist Intelligence Unit	2006	130 cities in the world
5	Dritter Großstadtvergleich	IW Consult GmbH/Institute of the German Industry	2006	50 German cities
6	Europas Attaktivstes Metropolen für Manger	University of Mannheim/Manager Magazin	2005	58 European cities
7	Les Villes Européennes: analyse comparative	UMR Espace (Rozenbiat, Cicille)	2003	180 western European cities

the weights in some evaluations (such as 1) is based on the results of interviews of target cities.

The team of Prof. Giffinger built intelligent city evaluation indicator system frame model first based on preliminary theories and empirical researches. The first version of intelligent city evaluation indicator system model uses hierarchical indicators, including 6 characteristics, 31 factors and 74 quantifiable indicators, to sort and research medium-sized cities in Europe through this model (see Fig. A.4).

6 intelligent city characteristics are intelligent economy, intelligent residents, intelligent governance, intelligent transportation, intelligent environment and intelligent life. Each characteristic has a number of factors respectively. Viewing from the characteristics and factors, some of them are hardware constructions that are more technical, such as equipment and facilities, and others are economic factors such as productivity levels, while more of them are evaluations for software, i.e. measurement on development level of social capital. Social capital is not only a hot point of academic discussion, but also an important soft power for city development. In order to evaluate such factors, it should further decompose intelligent city factors into quantifiable indicators.

For selection of research objects, it shall screen in 1595 cities determined by European Union related researches according to three constraint conditions. First, only medium-sized cities can be selected, where the population size is limited to 0.1–0.5 million. Second, there must be at least one university in the city as the basis of knowledge production. The last condition is excluding satellite cities of big cities (there is no large city with a population of more than 1.5 million nearby). In addition, the city must be within the scope of Urban Audit city database of EU. With the further constraints in the availability of data, 70 cities were screened out as research objects gradually.

The research established a research database for 70 cities and 74 indicators mainly using secondary data from research projects at EU level, and preformed

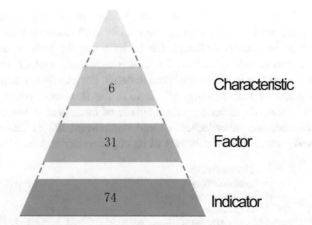

Fig. A.4 The first version of intelligent city evaluation indicator system model developed by TU Wien

standardized process for the data in order to realize indicator integration and horizontal comparison. The research acquired 6 characteristics and overall performance scores of 70 cities, then got corresponding ranking (see Fig. A.5) and distributed them on the map for investigation.

At same time, with the analysis the evaluation results combined with the actual situation of the city, we could set up the general situation of each intelligent city, and put forward intelligent city development strategic direction for each city on the basis of their own conditions and comparison.

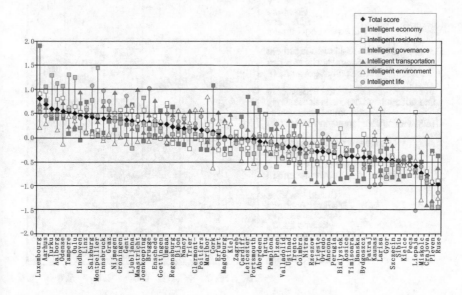

Fig. A.5 Score and ranking of the whole and characteristics of cities

The overall ranking is only to position. A city with higher ranking are not necessarily doing well in every aspect, while a city with lower ranking might have some prominent indicators. Although the balance may be poor, it also has distinctive development characteristics. So, in this evaluation system, strengths and weaknesses of each factor will be analyzed in detail. For example, when concluding factors of aspects of Luxembourg, which took the first spot, you can see that, flexibility and creativity under the characteristic of intelligent residents as well as education facilities and other indicators under intelligent life of Luxembourg are still very weak, which are short boards of its city development (see Fig. A.6).

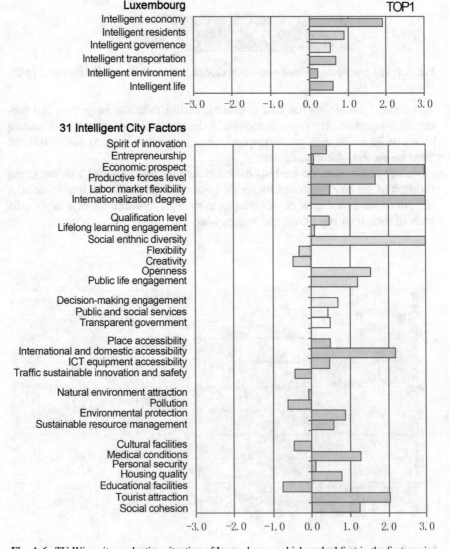

Fig. A.6 TU Wien city evaluation situation of Luxembourg, which ranked first in the first version

This evaluation system could not only evaluate and rank cities in a whole, but also perform specific analysis on a certain city, and even realize city evaluation and comparison through breaking it into indicators level in order to help identify specific problems of city. For example, only for the medical conditions factor under the characteristic of intelligent life, city distribution is analyzed based on secondary indicators to determine the development status of the cities and development characteristics of regional spaces.

2. The Second Version of the Evaluation Indicator System: Improvement on Methods and Visualization

Prof. Giffinger's team has performed the research of the second version of intelligent city evaluation indicator system from the end of 2012 to the beginning of 2013, the result of which was published in 2013. The second version updated the database and ranking based on the original intelligent city model. It enlarged the number of cities to 71 on the basis of intelligent development and data availability.

Under the frame of original evaluation model method, the second version has changed the original factors to domains, adjusted the number of domains to 28, replaced indicators with components and adjusted its number to 82 according to changes of city development and evaluation requirements while keeping the main frame stable (see Fig. A.7).

In 2013, the model also introduced new intelligent city evaluation overview function. It could directly select several cities from database to perform horizontal comparison on the basis of the discussion on city's specific characteristics and factors. Intuitive comparison of six characteristics of the cities ranked 1st (Luxembourg), 2nd (Aarhus) and 13th (Graz) in 2013 is shown as Fig. A.8.

Meanwhile, the indicator data of each city were described by city overview, including specific evaluation values of 6 characteristics and evaluation values of

Fig. A.7 The second version of intelligent city evaluation indicator system model developed by TU Wien

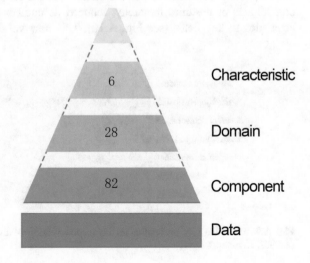

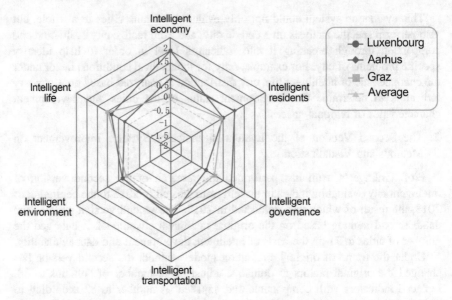

Fig. A.8 TU Wien city's horizontal comparison tool of the second version of the evaluation indicator system

domains of sub-hierarchy. Additional visualization functions and an open self-service data platform have made evaluation methods and research processes more open and transparent. The evaluation overview of Aarhus, Denmark, which ranked 2nd in the second version, is shown as Fig. A.9.

3. The Third Version of the Evaluation Indicator System: Wider Application

The third version of intelligent city evaluation indicator system published in 2014 basically continued to use the methods in the second version, fine-tuned the components of the third hierarchy, reduced its number to 81, and changed characteristics to key fields (see Fig. A.10). This new version is supported by 7 EU

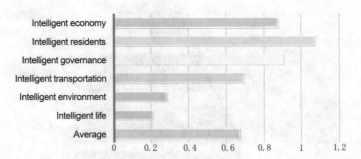

Fig. A.9 TU Wien city evaluation in the second version of the evaluation indicator system (Aarhus, Denmark)

Fig. A.10 The third version
of intelligent city evaluation
indicator system model
developed by TU Wien

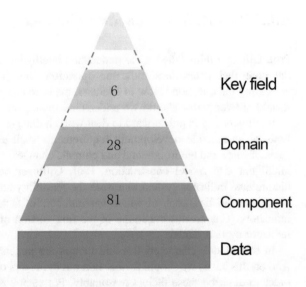

Seventh Framework Programme and Planning for Energy Efficient Cities, PLEEC, at the same time (Giffinger et al. 2014a, b). Due to the requirements of PLEEC project research, all of 6 case cities cooperating with the project were included as evaluation objects on the basis of the original, so the number of evaluation objective city is added to 77. Because 2 of the added cities are not covered by Urban Audit database, the immediate data will be collected to perform research.

The third version continued to use the visualization and display technologies in the second version, and updated the data. At the same time, due to the research requirements of PLEEC project, it performed energy intelligent city research on 6 case cities in respect of energy efficiency. It performed evaluation in a similar hierarchical method on the basis of data of two investigations and about 100 interviews, which formed *Energy Smart City Profiles*. The third version selected a subsystem section of intelligent city, designated domains and components and determined the weights according to immediate investigation data. Then, it performed more detailed description and comparative research through second investigation and put forward targeted development direction and recommendations in terms of intelligent city construction path for promoting sustainable energy for cities.

The third version of intelligent city evaluation indicator system paid more attention to specific objects to describe and understand intelligent city overview through the integration with PLEEC project requirements. Intelligent city is a general concept, so when actually evaluating cities, we should combine it with specific environment.

A8.2 Research Review and Outlook

Prof. Giffinger thinks the way to understand intelligent city and city development is the core that determines indicator operation and direction. Therefore, before answering the question "How to evaluate, measure and rank intelligent cities?", we should answer "How should we understand intelligent city?"

Intelligent city is put forward to deal with challenges that cities are facing on the basis of sustainable development requirements such as economy reconstruction, social change and environmental and climatic changes. In the normative research of intelligent city model construction, Prof. Giffinger determined the structure of hierarchical indicator system to ensure the feasibility and put forward 6 intelligent city characteristics, each of which has numbers of factors and consists of several indicators. This creates the frame of the first version of intelligent city evaluation indicator system model.

In this model, characteristics and factors are understanding of intelligent city. And on this basis, we should further find out indicators or indicator combinations in order to describe these factors reasonably. For each basic indicator evaluating the intelligent city, Prof. Giffinger thinks due to the close relationship in city networks, indicator selection should not be limited to indicators of the city itself. For the selection of some indicators, we need to consider indicators of a lager spatial dimension. For example, some regional even national indicators could also be used to measure city performance.

The evaluation rankings of three versions are based on a same intelligent city model and same understanding of intelligent city. The latter two versions replaced factors with domains in order to describe the information contents covered by them more precisely. In addition, during the research in cooperation with PLEEC project, when empirical research on 6 case cities was performe, it's found that it's difficult to describe the status of each case city precisely or carry out horizontal comparison on these cities only according to the statistical data. The first version indicators are set according to EU Urban Audit city database, and the data of indicators are required to be precise and specific. Therefore, the case city not covered by the database could not provide all the corresponding data.

In 2014, Prof. Giffinger put forward an idea that under the original model frame, the specific indicators of level three should not only rely on quantifiable statistics indicators, but also need to be addressed to various stakeholders in the city (such as government, enterprises and citizens). Qualitative research methods were partly adopted. Therefore, during the research in cooperation with PLEEC project, for 6 case cities, a set of survey questionnaire was designed to make sure the degree of importance of all domains under different key characteristics, and further discussed about which domains have higher recognition degree, as well as what aspects the relevant parties want to promote. Prof. Giffinger noted that, although interviews and researches for 6 case cities have taken so much time and energy, domains and even components and their weights in the evaluation system are still sensitive, and may be different depending on relevant parties, time and population discussed for a certain case city but have relative stability in general. Cities have been in a dynamic process, and intelligent cities should have local economy, culture, and environment and time characteristics of their own. Prof. Giffinger thinks that such evaluation system and information data acquisition methods more suitable for the requirements of this target characteristics of cities.

Although the intelligent city ranking is just one of the popularized research results, the indicators and methods behind it are based on the understanding of cities and their intelligence. Prof. Giffinger thinks intelligent city needs social innovation methods to guide and support the development of technologies. This is a self-learning process of cities and should not only be guided by technologies. For evaluation indicators and especially research methods which determines weights, the three versions of intelligent city evaluation model have changed from relying on unified statistical data to expressing ideas of parties of cities. So to speak, its own intelligence has been promoted through self-learning and improving. City evaluation and ranking should be regarded as a toolkit for experience learning, problem diagnosis and policy adjustment of cities, which plays a practical role in the formulation of city's development strategies.

A8.3 Application and Popularization

Since the first version of intelligent city evaluation report was published in 2007, the research has been carrying out in Regional Research Center in Vienna University of Technology. In addition to the discussion and research improvement in the field of academic (such as PLEEC research project), contents related to the research are popularized and applied in intelligent city practices in Europe and even the whole world to some extent. In Austria, some cities such as Vienna, Graz and Linz, the results of the evaluation method were used to formulate intelligent city development strategies. In the scope of Europe, Ljubljana of Slovenia, Bilbao of Spain, Krakow of Poland and some other cities also use the research method to guide the development of intelligent cities. At the same time, Prof. Giffinger also provides advices for cities in Germany, Israel, Japan and other countries based on intelligent city evaluation and development as a consultant.

Bilbao continues to use above intelligent city model to perform its intelligent city research, and on the basis of this, they also put forward corresponding indicators as the development guide for intelligent cities based on case research and its own development characteristics.

Then intelligent city project of Krakow—SMART_KOM was started in 2013. The team led by Prof. Giffinger was responsible for the formulation of intelligent city strategies as an international participant of the project. Krakow Science and Technology Park was the organizer of the project, but in the strategic level, they cooperated with Krakow and Krakow Metropolitan Area in a higher space level.

The project was conducted in two phases (see Fig. A.11). The project content was divided into three pieces: problem diagnosis, case researches integration, and

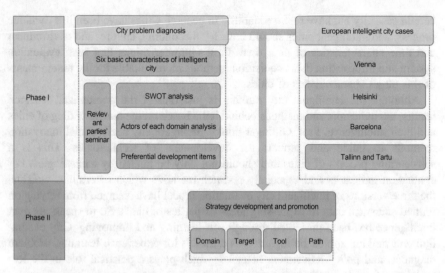

Fig. A.11 Krakow intelligent city project—SMART_KOM promotion frame

strategy development and promotion. The first two were conducted at the same time as the basis. At first, under above intelligent city frame model, they discussed on 6 basic characteristics—intelligent residents, intelligent life, intelligent environment, intelligent economy, intelligent transportation and intelligent governance respectively. From November, 2013 to March, 2014, they organized work seminars—Smart City Work shop, which attracted 161 participants in total. Before each work seminar, they prepared basic reports. Each working group would prepare preliminary reports, collect status and data of each space level under the topic of the seminar, and perform basic analysis as primary diagnosis. On the seminar, they would perform SWOT (strengths, weaknesses, opportunities and threat) analysis for a certain intelligent domain, and put forward city problems that require intelligent solutions most. Then, each party put forward relevant actors of particular intelligent city sub-systems.

At last, they would determine preferential actions in each domain through discussion and consider key development points as behavior and implementing goals.

On the basis of 6 topic seminars, an intelligent city comprehensive seminar integrating results of the 6 discussions and a joint seminar of regional government's functional departments were held in April, 2014. The 6 seminar has collected many ideas and future visions for Krakow and regional intelligent city development potential, and set some of preferential development items.

Of course, these preferential development items were obtained through specific group discussion, which were coordinated according to the overall development needs. In next phase, they collected and researched further data to put forward regional, city and park's development strategies based on the model setting. Vienna University of Technology also continued to play an important role in this as the scientific research team of city planning background. Prof. Giffinger think, in the

Krakow intelligent city project, because we need to consider the development requirements of three space levels, quantitative indicators has lost its most significance. In the level of city, they could be compared under the whole European background, but in the level of region and considering the coordination of city development, this project used qualitative analysis and research methods, i.e. determining preferential development items (not indicators) of domains under characteristics, as well as relevant actors specific to items, and set corresponding development goals of some items. In the next phase, we're looking forward to formulate further intelligent city development strategies on the basis of these preliminary works to realize a smooth evolution from theory to practice.

References

Cai DF (2012) About public security of cities. Democracy (2):8–9

Dong Ji C (2010) Introductory medical informatics. People's Medical Publishing House, Beijing

Giffinger R, Haindlmaier G, Kramar H et al (2014a) PLEEC report: energy smart city profiles [R/OL]. http://www.pleecproject.eu/downloads/Reports/Work%20Package%202/wp2_d23_energy_smart_city_profiles.pdf. Accessed 11 June 2016

Giffinger R, Haindlmaier G, Hemis H et al (2014b) PLEEC report: methodology for monitoring. http://www.pleecproject.eu/downloads/Reports/Work%20Package%202/wp2_d24_methodolgy_for_monitoring.pdf. Accessed 10 June 2015

Sheng XF, Yang XM (2012) Six dimensions of Pudong, "feel the pulse" of intelligent city. China Informatization (14):20–23

Yan YJ (2006) The characteristics and enlightenment of city grid management. City Probl (2):76–79

Zhao JY (2009) E-commerce industry highlights the logistics requirements. Logistics Mater Handling 14(10):43–46

Printed in the United States
By Bookmasters